American Submarine War Patrol Reports:
U.S.S. Wahoo (SS-238)

American Submarine War Patrol Reports: U.S.S. Wahoo (SS-238)

Riverdale Electronic Books
Riverdale, Georgia

American Submarine War Patrol Reports: U.S.S. Wahoo (SS-238)
Introduction and other new material © 2003, Riverdale Electronic Books. All rights reserved.

This work contains both copyrighted and public domain materials. For information, contact the publisher.

Riverdale Electronic Books
P.O. Box 962085
Riverdale, GA 30296

ISBN: 1-932606-01-7

Printed in the United States of America

Contents

Introduction	vii
Background	1
Patrol One, 23 Aug 1942 – 17 Oct 1942	2
Patrol Two, 8 Nov 1942 – 26 Dec 1942	20
Patrol Three, 16 Jan 1943 – 7 Feb 1943	36
Patrol Four, 23 Feb 1943 – 6 Apr 1943	52
Patrol Five, 25 Apr 1943 – 21 May 1943	80
Patrol Six, 8 Aug 1943 – 29 Aug 1943	94
Patrol Seven, 9 Sep 1943 – 11 Oct 1943	116
Appendix I, Navy Time Zones	119
Appendix II, Ship Type Designators	121
Appendix III, Sinkings Credited, Wartime vs. JANAC	123

Introduction

The American fleet submarine was arguably the most successful naval weapon of the Second World War. While never constituting more than 2% of total American naval forces, submarines were responsible for sinking more than half of all Japanese ships destroyed during the war. Nor was the true potential ever fully realized, for American subs spent the first 20 months of the war handicapped by torpedoes that suffered from multiple problems, which had to found and cured one by one, as they tended to mask each other.

Riverdale Electronic Books is now engaged in publishing the full set of American submarine patrol reports. Ultimately, we plan to publish every existing report. Some, obviously, do not exist, for some boats were lost on their first war patrol. Likewise, in this first volume in the series, which contains the reports filed by the two wartime commanders of U.S.S. *Wahoo* (SS-238), perhaps the most famous of all American fleet boats in World War II, the final report was never filed as the boat was lost with all hands on 11 October 1943, during her seventh war patrol.

While we have edited these patrol reports, we have mostly restricted ourselves to correcting obvious spelling errors, missing words (where the omission is easily recognized), and the addition of conventional typographic elements and symbols (such as degree symbols) that were not available in the typewritten originals. We have also taken some effort to improve readability. The originals were typewritten, but this edition is conventionally typeset and does not claim to be a facsimile edition. Words appearing in square brackets [ ] have been added by the editor, and indicate words that were probably unintentionally omitted from the original reports. (Or, at least, the editor's best guess.)

We have also added footnotes where they seemed warranted. Appendices provide information on the Navy's time-zone system, which used letters to indicate, ship types, and a comparison of wartime and JANAC sinking credits.

Perhaps needless to say—though we will—the classification indicia from the originals no longer apply, these reports having long since been declassified.

Background

U.S.S. *Wahoo* (SS-238) was laid down on 28 June 1941, at the Mare Island Navy Yard in Vallejo, California. She was launched on 14 February 1942, under the sponsorship of Mrs. William C. Barker, Jr., and commissioned on 15 May 1942. Her first commander was Lieutenant Commander Marvin G. Kennedy, a member of the Annapolis Class of 1929.

A *Gato* class fleet submarine, *Wahoo* was 311' 10" in length, with a 27' 4" beam and a mean draft of 15' 2". As built, she was armed with a single 3"/50 deck gun, two .30-caliber and two .50-caliber machine-guns, and ten 21" torpedo tubes. The torpedo tubes were allocated six forward and four aft.

Wahoo did initial training along the California coast, traveling as far south as San Diego. This completed, she departed San Francisco on 12 August, and arrived at Pearl Harbor six days later. *Wahoo* then participated in training exercises until 21 August. At that time she returned to Pearl Harbor to make final preparations for her first war patrol, which commenced two days later.

Patrol One, 23 Aug 1942 – 17 Oct 1942

CONFIDENTIAL

Subject: U.S.S. *Wahoo*—Report of First War Patrol. Period from August 23 to October 17, 1942.

Area: North of latitude 7°-15' and SW of a line bearing 315°T from Tauarar Pass.

OPERATION ORDER: Number 73-42.

1. NARRATIVE.

August 23—09:00 (V-W) Departed Pearl. Made trim drive and received two indoctrinational depth charges from escort. Set speed 14 knots for area.

August 26—06:30 (X)* Made trim drive

August 26—22:30 (Y) Crossed 180th Meridian.

August 28—07:00 (M) Lookout reported sighting airplane. Unconfirmed. Improbable.

August 29—07:20 (M) Submerged. Surfaced at 17:30.

August 29—18:30 (M) Aircraft contact on Radar at ten miles. Sky heavily overcast and squally. Submerged. Surfaced at 1900. Position ninety miles East of Taongi Atoll.

August 31—06:25 (L) Sighted airplane about six miles ahead. Submerged. Surfaced at 1730(L).

---

* Time is given in military (24-hour) time. The letters in parentheses following the time refer to the Navy time zone system. (See Appendix I.)

2

U.S.S. *Wahoo* (SS-238)

September 1—06:00 (L) Aircraft contact on Radar at three miles. Submerged. The time of day and geographical position (north of Ponape) made this contact most unexpected. However there was a very bright moon and possibly this area is being given a thorough search. During the night O.O.D. had picked up three greenish lights at about half hour intervals in the distance. There may be a connection between the lights and the early morning contact.

September 1—Sound contact ahead. Nothing in sight by periscope. Sound followed contact 30° change in bearing and reported propeller beats and speed changes. Lost contact after about twenty minutes. (07:20(L).

September 3—05:00 (K) Commenced periscope patrol. Submerged outside of assigned area between Hall and Namonuito Islands about 40 miles NE of boundary line, proceeding southwesterly.

September 4—23:00 (Y) Sighted Tol Island bearing 090° T, distance about 30 miles.

September 5—04:30 (K) Picked up fast propellers, which crossed the bow heading for Truk. With a bright moon and glassy sea nothing was sighted, so it was presumed to be a small patrol boat.

September 6—05:25 (K) Sighted a loaded ship similar to *Wyogo Maru* class tanker, S77, except there was only one stack. Approached and fired three torpedoes at range of 1430 yards (using 50ft. masthead height). Position 15 miles off Truk. Target headed for Piaanu Pass. Observed airplane on far side of target at about three miles just before firing. Observed target for about one and one half minutes after firing, and saw torpedoes leading her track. About 1-1/2 minutes after firing she started a turn toward us. Assured the torpedoes had all missed or run under so turned toward and went deeper to run under target. At approximately 2 minutes, 20 seconds after first shot one explosion was heard followed shortly thereafter by a second explosion. Lost sound of propellers shortly after explosions. From the timing it would appear that the actual firing range was about 3,000 yards and that at least one and probably two torpedoes were detonated by the target or its wake. Because of sea conditions, proximity of the base, and the presence of an airplane screen no periscope observations were made after the explosions. There is no visual evidence to warrant claim of damage to the enemy.

September 8—07:48 (K) Sighted reconnaissance Bomber type 95 at about two miles distance.

September 10—06:12 (K) Heard distant underwater explosion, followed by others at 06:20, 06:34, 06:44, and 06:53.

September 10—13:45 (K) Heard five (5) distant underwater explosions at about ten (10) seconds apart.

September 11—00:10 (K) Picked up propellers bearing north, and lost sound contact about ten minutes later on NW bearing. Nothing sighted. Presumed it was a small patrol boat passing fairly close aboard.

U.S.S. *Wahoo* (SS-238)

September 13—04:52 (K) Heard two (2) underwater explosions about ten (10) seconds apart followed by two (2) more at 04:55. Estimated distance of 8 to 10 miles.

September 14—10:25 (K) Picked up propeller noises, and shortly thereafter sighted ship bearing 270°T, distance 12,000 yards, angle on the bow 65° port. Ship was screened by a single floatplane similar to Navy Reconnaissance Bomber type 95, and by a small patrol boat. Came to normal approach course and approached by sound for thirty-two minutes. At the end of thirty-two minutes came to periscope depth for the shot and found range to be 4,000 yards and track about 110 port. Target was identified as a small freighter of about 2,500 tons, and being in an unfavorable firing position broke off the attack. Ship proceeded into Truk via Piaanu Pass.

The surface escort was not observed through the periscope, but was plainly heard and tracked by our sound. It was the same high-pitched, fast propeller beat as was heard on the 5th and 11th of Sept. The escort would run for about five minutes and get ahead of the freighter, then lie to and drop astern. It was presumed that he had listening gear but no echo ranging. There was no indication that we were detected, which is further born out by the fact that our target made a 25° zig in our direction during our approach.

September 14—20:55 (K) While engaged in routine servicing of torpedoes No.1 tube was inadvertently fired with a fully ready war shot in the tube, both outer and inner doors being closed, and with 400 lbs. per square inch of impulse air. Details are given under "Major Defects Experienced." This tube is out of commission for the remainder of the patrol.

September 20— Having spent seventeen days in Southeastern part of the area and seeing but two small ships heading toward Piaanu Pass, decided to spend a week patrolling south of Namonuito Island for East-West traffic.

September 20—22:55 (K) With a bright moon, clear sky, no wind, and a flat sea, the O.O.D. sighted a column of smoke bearing 322°T. Ran toward it for half an hour on the surface and then submerged for a periscope attack. At 00:01 target was identified as a freighter of about 6,500 tons of the *Keiyo Maru* Class, similar to that of plate 69. Course estimated at 135dT. Speed 12 knots at this time, although at various times the ship would stop and lie to for appreciable periods.

September 20— It was later discovered that we were at her rendezvous point with an escort. At 00:05 it was seen that our tracks were very close together and we swung left for a stern tube shot on the starboard track. Target passed abeam about 200 yards distance and at 00:08 we fired the first torpedo on 140 starboard track. This torpedo undoubtedly failed to arm. The second torpedo was fired with a 2° right spread on 155 starboard track. It apparently ran down the starboard side of the target and target saw it and turned left. Third torpedo was fired with a 2° left spread on 162 starboard track and also missed. By this time target presented a 90° port angle on the bow, and

U.S.S. *Wahoo* (SS-238)

with a fourth new setup on TDC another torpedo was fired on 108 port track. This torpedo hit the target 1 minute and twenty seconds after being fired. Target took a port list of about 50° and settled bodily and by the stern, as witnessed by four or five observers, and her engines slowed radically. Four minutes later there started three series of underwater explosions, each series consisting of three or four explosions, and when upon observation through the periscope we first sighted the escort arriving on the scene. He dropped perhaps a dozen depth charges, none within a thousand yards of us.

At about 00:30 there was considerable confusion around the target. Accompanied by sounds believed to be internal explosions, a billow of heavy black smoke came off the target, and it was no longer seen or heard. By this time we were about 4,000 yards away so we surfaced to clear the area.

About the time we got to 21 knots the escort picked us up and with a tremendous stream of black smoke pouring out of his stack he gave us chase. He was closing the range by leaps and bounds when along came the most welcome rainsquall that it has been our pleasure to encounter. We passed into the squall and made a radical course change. When we came out the other side we found that the escort had followed us through so we eased back into the rain and went the other way. When next we got into the clear the escort was barely visible on the horizon and he did not pick us up again.

There is no doubt in the minds of any of this crew but that we sunk the freighter.

September 24—**22:00** (K) Sighted patrol boat bearing 090°T., distance 5 miles, angle on the bow zero. Moon bright, sea calm, visibility good. Submerged. Patrol passed about 3,000 yards astern, heading west. Surfaced at 23:30 and remained in area to see if he would return escorting a target. Such was not the case.

September 25—**04:15** (K) On routine daylight dive there was some trouble in getting the ship down. Upon surfacing at night found that the Bow Buoyancy vents were not operating properly. Removed bow buoyancy manhole cover for remainder of patrol.

September 30—**05:20** (K) Sighted ship later identified as Aircraft Tender *Chiyoda* bearing 322°T., course 170°T, range 12,000 yards. Estimated target was headed for Piaanu Pass and that we were in ideal attack position. Two minutes later target zigged 40° left. Three minutes after this zig he went 35° further left. We turned to normal approach course and closed the range to 6,000 yards, at which point we were on 130° starboard track. Target then made another left zig presenting 175 starboard angle on the bow and went over the hill on course 075°T., heading apparently for North Pass in to Truk. Weather was ideal for submarine approach and we were able to watch target continuously except when making high speeds to close the range. There were no screens or escorts, and any planes [that] might have been overhead never came within the periscope field. The Japs were just begging someone to knock off this Tender, but it was not our lucky day. In 24 minutes he zigged 95° away from us.

September 30—**07:45** (K) Sighted airplane to Northward flying east. 08:50(K) Sighted airplane to Northward flying east.

October 1—**07:00** (K) Sighted airplane to Northward flying east.

October 1—**1412** (K) Sighted smoke bearing 220°T. Smoke in sight for two hours. By plot we estimate it to have been a ship on Easterly course, speed 10 to 15 knots, minimum distance to our track 18 miles.

October 1— From what we have seen and heard it is believed that a great portion of the shipping from the Empire to Truk is running close to Namonuito, with medium and large ships entering North Pass. We cannot patrol pass between Namonuito and East Fayu so, having but a few days left, decided to patrol for ships making a landfall on Ulul. This lane is not in any assigned submarine area. It is believed that the spirit of our orders will permit the stretching of our Western patrol limit to Longitude 149°E., especially if events turn out to favor us with a target or to discover a main shipping lane.

October 3—**13:50** (K) Sighted Ulul Island, Namonuito, bearing 035°T., distance 9 miles.

October 3—**15:35** (K) Sighted fishing boat bearing 122°T., on course 320°T., distance 6,000 yards. Boat passed 1,000 yards abeam and continued on to North West. While watching it sighted masts of a similar boat to Northward. We were at this time about 8 miles West of Ulul Island. First boat was in sight about an hour, the second was only seen once.

October 4—**19:20** (K) Sighted lights of a small craft bearing 220°T., and lost them 20 minutes later on bearing 244°T. Estimate it to have been a fishing boat similar to one seen on October 3, on course NW., and passing about 4 miles south of our track. There were two white lights in a horizontal line about 50 feet apart, and periodically a red flashing light showed between and below the white lights. We were on the ship's starboard beam, so it was not a sidelight. Believe it to be a station ship for incoming traffic. Remained in position, lying to awaiting return of small craft in company with suitable target.

October 5—**04:00** (K) Received ComSubPac dispatch assigning us to Southern Sector of area in addition to sector already assigned. Started South and East to head for new sector.

October 5—**06:54** (K) Sighted aircraft carrier *Ryujo* accompanied by two *Amagiri* class destroyers bearing 220°T., angle on the bow 60° starboard, range 11,000 yards, speed 14 knots. One DD was leading and second was trailing carrier. Made approach which, upon final analysis, lacked aggressiveness and skill, and closed range to about 7,000 yards. Watched the best target we could ever hope to find go over the hill untouched at 08:00. A normal approach course at time of sighting and full speed for the whole twenty minutes would have brought us in to 3,000 yards and a fair shot. At 09:15 we surfaced and went ahead at 19 knots on course North, which was the target's last course, in an endeavor to trail. About 10:30 we ran into general rain squalls and reduced visibility and at 12:00 broke off the chase. At 13:15(K) sent message to ComSubPac giving contact. Called on 16460 KCs[*] for 5 minutes, received no answer, so broadcast message twice. Considerable interference from Japanese station.

October 6—02:30 (K) Having opened to SW to change R.D.F. bearing from Truk, and not receiving any indication on Submarine Radio schedule that our contact report was received, sent report again on 4235 and 8470. Called for 20 minutes, then transmitted message twice. There was considerable interference from the Japanese and no report was obtained.

October 7—12:00 (K) Departed Patrol area.

October 8—01:20 (K) Having made three tries at getting off our message without success, using all the high-priced help and mechanical aids, we felt pretty discouraged and had ordered the usual radio silence. The radioman on watch, on his own initiative, tuned in on 4235 KCs and listened. He states that suddenly reception on this circuit became exceptionally strong, so he grabbed the message, called Pearl, sent the message and got a receipt. He then reported his violation of orders. Held mast and gave him a reprimand for his offense, and advanced him one grade in rating for his loyalty, initiative, and ability to get results.

October 7—20:50 (K) Broadcast contact report twice on 4265 KCs without interference, then called Pearl and Midway for about 30 minutes. No answer.

October 9—12:40 (K) Sighted airplane. Submerged. Surfaced at 16:00.

October 10—17:00 (K) Aircraft contact on Radar at 2 miles. Sky almost completely overcast. Submerged.

October 12—12:20 (L) With the sky overcast and squally, sea rough with lots of whitecaps, sighted a Mitsubishi 97 two engine, Army heavy bomber close aboard and headed directly for us. His range was less than 1 mile, position angle 30°. We made a pretty fast dive. There was a long silence. The only reason we can attribute to the lack of attacks is that he was either unarmed or was as surprised as we were. There was no indication on our Radar. Counting the O.O.D. there were seven (7) lookouts on the bridge when the plane was sighted.

October 15—09:45 (X) Sighted U.S.N. PBY at about 10 miles. He passed within about 3 miles without registering on the Radar. Perhaps the SD Radar is not functioning.

---

[*] KC – Kilocycle, an older measure of radio frequency. The term Kilohertz has been commonly used since the 1960s.

U.S.S. *Wahoo* (SS-238)

October 16—15:00 (WX) Sighted U.S.N. PBY and exchanged recognition. Radar inoperative.
October 16—17:50 (VW) Sighted U.S.N. PBY flying E, altitude 1,600 feet.
October 17—06:30 (VW) Met escort at rendezvous and proceeded to Pearl.

2. WEATHER

ENROUTE TO TRUK: Normal trade weather. Sea from NE or E, condition 1 to 3. Trade winds to Eniwetok, then shifting southerly. Sky generally overcast with occasional rain.
ON STATION – TRUK: Glassy sea with no swells or white caps. Clear sky with some low clouds on horizon, especially during night. On the 8th the sea picked up a bit to condition 1, and visibility decreased. Very little breeze at night, humidity high. During the 9th the wind died down and the sea again became flat and glassy. Rainy and overcast on the night of 10-11 September, with sea condition 1, and NW breeze. During the day the wind shifted to southerly and sea became good for periscope observations for the first time. By the 12th the sea was condition 2, with white caps. Sky overcast and visibility spotty, but where clear of rain squalls horizon visible for 15-20 miles. Winds remained variable in strength and direction. During later part of September there was considerable rain, with shifting winds and seas, alternated with short periods of flat calm. Rain squalls covered large portions of the horizon, with visibility greatly reduced. This was no particular handicap until we tried to follow the *Ryujo* on the surface. We ran afoul of an area of reduced visibility that morning entered it at 10:30, and remained in it until the chase was abandoned at 12:00. During this period the visibility was variable between 100 and 3,000 yards.
ENROUTE TO PEARL: The sky was almost completely overcast with much rain and reduced visibility the entire trip. From Truk to Taongi the sea was calm, with slight swells from S and SE. After passing Taongi the sea shifted to E and became moderately rough, condition 3 to 5. Normal trade winds were encountered north of Latitude 15°N.

3. TIDAL INFORMATION.

The normal set was in the direction of the prevailing wind, generally toward NW with drift from 0.4-0.7 knots. In the passage between Namonuito and Hitchfield-Gray Feather Banks and inshore west of Truk strong easterly sets were encountered regardless of the wind and surface condition of the sea.

U.S.S. *Wahoo* (SS-238)

4. NAVIGATIONAL AIDS.

Tol Island, Truk was in sight a good portion of the time on station. Ordinarily the peak could be sighted from 35 to 40 miles, and was very valuable as an aid in checking position. Ulul Island, Namonuito, was sighted about 9 miles off and is an excellent landmark.

No other Navigational Aids were noted.

5. ENEMY SHIPS SIGHTED:

Time & Date	Position	Course	Speed	Description	Remarks
05:25 (K) 6 Sept	15 Miles WNW of Truk enroute Piaanu Pass	145°	10	2400 ton tanker loaded, similar to that of plate 577 except with only one stack. One gun about 3" mounted at bow.	Escorted by one airplane. Hit by at least two torpedoes.
10:15 (K) 14 Sept	12 Miles W of Piaanu Pass	135° - 110°	10	2,500-ton freighter similar to *Sensei Maru* listed on plate 163.	Escorted by one airplane and surface patrol.
22:25 (K) 20 Sept	Lat. 7°-43' N Long.150°-36' E	135°	12	6,500-ton freighter similar to *Keiyo Maru* listed on plate 69.	Passed about 200 yards abeam during bright moonlight. Sunk.
00:10 (K) 21 Sept	Lat. 7°-43' N Long. 150°-24' E.	—	25	Single stack ship of about 700 tons oil burning. 163.	Seen at about 4,000 yards in moonlight. This was the escort of the above listed ship.
22:00 (K) 24 Sept	Lat. 8°-30' N Long. 150°-02' E	270°	12	Patrol boat similar to our old 110 ft S/M chasers. Very long and low with single stack and bridge structure slightly forward.	
05:45 (K) 30 Sept	Lat. 7°-58' N Long. 151°-02' E	170° - 075°	18	Aircraft tender *Chiyoda*. No planes on deck.	Unscreened. Headed for North Pass, Truk. Five aircraft seen in area during this period.

15:35 (K) 3 Oct	8 miles west of Ulul Island Namonuito	330°	8	Fishing boat of about 150 tons. Rounded bow, elevated forecastle, 2 masts with fore about 50 ft. and main about 30 ft. high. Two booms rigged from amidships at angle of about 45°.	No armament noted.
06:54 (K) 5 Oct	Lat. 9°-15' N Long. 149°-00' E	345° - 000°	14	Aircraft carrier *Ryujo* and two *Amagiri* class destroyers.	No aircraft on flight deck. All four masts up.

6. Description of all aircraft sighted, including type, position, course, altitude and time of sighting.

Time & Date	Position	Type	Course	Altitude	Remarks
15:00 (VW) 23 Aug	Lat. 20°-50' N Long. 159°-35W	USN PBY	080°	2000 feet	
15:10 (VW)	Lat. 20°-35' N Long. 159°-35' W	USN PBY	080°	2000 feet	
18:30 (M) 29 Aug	Lat. 12°-30' N Long. 170°-05' E	Large plane	000°	3000 feet	Radar contact 100 miles east of Taongi
06:25 (L) 31 Aug	Lat. 12°-45' N Long. 161°-00' E	–	East	2000 feet	Sighted at long range, position 60 miles north of Eniwetok.
06:00 (L) 1 Sept	Lat. 11°-30' N Long. 158°-00' E	–	–	–	Radar contact.
05:25 (K) 6 Sept	15 miles WNW of Truk.	–	East	2000 feet	Sighted while making approach. Did not observe closely.
07:50 (K) 8 Sept	40 miles WNW of Truk.	Reconnaissance bomber type 95 single float biplane.	NE	2000 feet	
10:25 (K) 14 Sept	12 miles W of Piaanu Pass, Truk	Small, fast planes	East	1000 feet	Air screen for freighter.
05:20 (K) 30 Sept	45 miles WNW of Truk.	Small, fast planes	West	4000 feet	Formation flight observed at long range.
07:45 (K) 30 Sept	40 miles WNW of Truk.	Single float seaplanes (95)	East	2000 feet	

U.S.S. *Wahoo* (SS-238)

08:50 (K) 30 Sept	35 miles WNW of Truk.	Single float seaplanes (95)	East	2000 feet	
07:00 (K) 1 Oct	35 miles WNW of Truk.	Small seaplane.	East	2000 feet	
12:40 (K) 9 Oct	Lat. 12°-35' N Long. 158°-50' E	Unidentified	East	3000 feet	Sighted at long range.
07:00 (L) 10 Oct	Lat. 14°-00' N Long. 162°-30' E	–	–	–	Radar contact at two miles.
12:20 (K) 11 Oct	Lat. 15°-50' N Long. 171°-34' E	Two motored bomber similar to Mitsubishi 97 Army heavy bomber	South west	3000-4000 feet	
09:45 (X) 15 Oct	Lat. 19°-20' N Long. 168°-00' E	USN PBY	East	2000 feet	
15:00 (WX) 16 Oct	Lat. 20°-35' N Long. 161°-24' E	USN PBY	East	1200 feet	
17:50 (VW) 16 Oct	Lat. 20°-35' N Long. 161°-16.5' E	USN PBY	East	1600 feet	

7. SUMMARY OF S/M ATTACKS.

Listed on printed form herewith attached as enclosure (A).

8. ENEMY A/S MEASURES.

Radar contacts indicated an A/S air patrol in the vicinity of Truk during twilight hours. This apparently covered a distance of but a few miles beyond the reef.

There is apparently a periodical sweep of the area by an offshore A/S air patrol during daylight which is infrequent.

Ships are escorted during daylight by aircraft and by a surface patrol. One incoming ship made a rendezvous 60 miles West of Truk at midnight with an A/S patrol boat of about 700 tons.

Surface patrols of small craft pass through or patrol the area at infrequent intervals. It is possible that those heard and seen were either meeting, or had completed escorting, surface vessels.

The presence of an airplane in the area usually meant that surface vessels were in transit.

No echo ranging was heard during any part of the patrol.

U.S.S. *Wahoo* (SS-238)

We encountered only one patrol that actually stopped to listen. It was acting as an escort. We heard ships at ranges varying from 3,000 to 12,000 yards, but there was no indication that we were heard by sound.

Only one depth charge attack was encountered, and on that the A/S vessel dropped charges at random with none inside our 1,000-yard range.

The area north of Marshalls is patrolled by air. Indications are that planes encountered were from Eniwetok and possibly Taongi.

9. MAJOR DEFECTS EXPERIENCED.

1. Torpedo tube number 1.

Tubes 1, 2, and 3 had been made ready for firing during the approach on September 14, but were not fired. After surfacing the torpedoes were examined in rotation.

Two tubes in each end are habitually kept ready for firing except for opening the outer door and air master solenoid valve. Forward tubes 3 and 4 are the ready tubes. While 3 was being checked, number 1 was made ready as the standby tube. Number 3 was checked and preparations made to fire a 25-50 pound inboard slug. Instead of raising the firing interlock and firing number 3 by hand, this was inadvertently done on number 1 instead. There was quite a bang, and tube number 1 flooded.

The torpedo officer was lowered over the side for external examination. He found the shutter intact, but the outer door sprung about 1 1/2 inches.

This tube was pumped down and the inner door opened. The torpedo had traveled forward about 3 or 4 inches, and there was indications that the torpedo had started a hot run which was speedily terminated by the over speed trip.

Several hours were spent in attempting to back the torpedo out of the tube, without results. When a 1 1/2 ton chain fall failed to budge the torpedo the attempts were abandoned.

This piece of gross carelessness has cost the ship the use of this one tube, and probably the wreckage of one torpedo. The full extent of the damage cannot be determined until return to port.

2. Bow buoyancy operating linkage failed to function properly and the vents would not open. The manhole cover was removed. The cause of failure will be investigated and remedied on return to port.

3. HARDIE-TINE H.P. air compressor.
Trouble continues with the discharge valve discs. A total of seven (7) first stage discs were broken during the patrol.

4. The ship's Hull Exhaust Ventilation system does not remove battery gases from forward battery when charging at the finishing rate and ventilating inboard. This condition can be corrected by installing dampers in the exhaust ducts in the pump room and galley.

5. SJ Radar caused considerable difficulty. The turning gear is of faulty design and becomes almost impossible to turn. The antenna tuning was thrown out of adjustment early in the patrol and without a willing target it is impossible to re-tune. Probably the major source of trouble came from improper operation by unskilled personnel, the whole unit will require checking and tuning. A small turning motor would be a distinct asset.

10. COMMUNICATION.

Radio reception – Radio reception of the NPM Fox Schedules was very good and was complete. The 8 megacycles were used during [the night] almost exclusively on station. 12 megacycles were used during the day to and from station and occasionally during the early morning while on station. Low frequency reception depended on weather conditions. Low frequency was seriously interfered with by the SD Radar. High frequency suffered very little interference.

Interference was encountered from enemy stations when transmitting on 4235 series. No interference was encountered on 4265.

Last serial received OCHER 031911 of Oct.

Last serial sent ANDROID 140930 of Oct.

11. SOUND CONDITIONS AND DENSITY LAYERS.

In general, sound conditions varied from good to excellent. Propellers were picked up at ranges varying from 3,000 to 12,000 yards. At times there was a marked variation in sound intensity of a steady propeller beat, with sound fading in and out. It was being stopped and started.

There were some sounds heard which were not caused by ships, these taking the form of grunts, groans, and clicking, but they were the exception and not the rule. They were heard in far less quantity than experienced in some other patrol areas.

Temperature gradients were measured daily. The water temperature remained 85° F. to 150 ft., with a rare gradient, except that a layer of water of 86° F could be found of about 30 feet thickness at depth varying between 60 and 120 ft. Sometimes a change in density of the water could be felt through a change in the trim but not indicated by temperature change of the water.

The fading of propeller noises is believed to have been caused by skip distances. Under conditions of zero temperature gradient, there results a bending up of the sound beam. This causes the beam to stay near the surface and bounce and re-bounce off the underside of the sea's surface. A sort of "skip distance" at a projector depth of 68 feet could easily occur under such conditions.

In general, it is felt that the sound conditions in the area to the Northwest of Truk are favorable for submarine operations. Submarine listening conditions varied from good to excellent, while the positive velocity gradient (resulting from a zero temperature gradient) combined with the ever-present density layer at a convenient depth would offer a fairly safe sanctuary against listening or echo-ranging by surface craft.

2. HEALTH AND HABITABILITY

During the early part of the patrol colds were numerous. As soon as we became acclimated and took proper precautions these cleared. There were no serious illnesses. One man had a small Furuncle on his right buttocks, which required lancing, and one Electrician's Mate suffered a second degree burn on his hand and was temporarily blinded by flash when he drew an arc in pulling a fuse from the auxiliary power board on a hot circuit.

Habitability was good. The average submerged temperature was about 90°F., with humidity decreasing shortly after diving. The air conditioning units were operated at capacity and were just adequate. On several occasions one air conditioning unit had to be secured for repairs, and it was not very comfortable at this time.

13. Miles steamed en route to and from station.

Miles steamed en route to station 3,109.
Miles steamed en route from station 3,075.

14. FUEL OIL EXPENDED.

Fuel oil expended en route station - 28,633.
Average speed - - - - - - - - - - - 12.9 knots.
Fuel rate - - - - - - - - - - - - - 9.22 gallons/mile.

NOTE: Conditions were ideal. Clean bottom, smooth sea, wind and sea from astern, and set in direction of travel.

Fuel oil expended returning from station 37,052 gals.
Average speed - - - - - - - - - - - - - - 12 knots.
Fuel rate - - - - - - - - - - - - - - - - 12.05 gals. per mile.

Three fourths of the return trip was made against head winds and seas, with adverse set. Some three and four engine speeds were used.

15. FACTORS OF ENDURANCE REMAINING

Torpedoes ---------- 17 (one probably wrecked)
Fuel ------------ 16,295
Provisions ---------- 30 days.
Fresh water ---------- 1000 gallons. This is no longer a factor of endurance unless the stills break down or become dirty. We could have ended the patrol with 10,000 gallons of water if we had desired.
Personnel ----------- 10 days.

16. The patrol was ended by the time written in the operation order.

17. REMARKS.

From departure for patrol on August 23rd until about September 30th this was a routine patrol, marred only by the unfortunate accident to No. 1 torpedo tube on September 14th. During this period we sunk one medium freighter and got one or two hits on a small tanker. The remainder of the patrol was a fiasco. On the 30th we sighted the *Chiyoda*. Actually it was not possible to get within torpedo range, but that was the bad luck of initial position and subsequent target movements, which is an excuse but not a suitable result. Then I decided to patrol West of Namonuito and to try to find out where ships were coming from. After sighting various small craft, suddenly on the 5th there came onto view the *Ryujo* heading for the Empire. Had I but required a more rigorous and alert watch we might have picked her up sooner. Had I correctly estimated the situation and made a more aggressive approach we could have gotten in a shot. Had I taken up the surface chase without allowing over an hour to elapse we might not have lost the target. Had I continued the search through the rain squalls until dark we might have picked her up again. None of these happened and the second target proceeded unharmed. The rest of the time allotted to the patrol was spent in changing position and attempting to transmit the information on the contacts to the Task Force Commander.

In studying over both approaches I find that they each conform to my normal method of attack, and confronted with the same situations again the results would probably be identical.

18. The Loop antenna was not used on this patrol. On previous tests it has given excellent reception at depths to 58 feet, where fading begins. All our submerged patrolling was done at 62 feet or deeper, so there was no occasion to use this loop.

U.S.S. *Wahoo* (SS-238)

(Endorsements to first war patrol report)

FC5-10/A16-3(FB5-102) SUBMARINE SQUADRON TEN

Serial 043　　　　　　　　Care of Fleet Post Office,
　　　　　　　　　　　　　San Francisco, California,
　　　　　　　　　　　　　October 20, 1942.

CONFIDENTIAL

From: The Commander Submarine Squadron Ten.
To: The Commander Submarine Force, Pacific Fleet.

Subject:　U.S.S. *Wahoo* (SS-238) – First War Patrol – Comments on

　　1. The *Wahoo*'s patrol covered a total of fifty-five days, of which thirty-four were spent on station. The last four days were spent patrolling to the westward of the assigned area. More time should have been spent in closer proximity to Piaanu Pass.
　　2. In the attack on September 6, there is insufficient evidence to support a belief that a hit was obtained upon the tanker. The masthead height was probably underestimated.
　　3. On September 14, sea and weather conditions are not stated. Under certain conditions, when an aircraft screen is known to be present, the practice of going deep between periscope observations is sound. However, to resort to an approach wherein no periscope observation is made for thirty-two minutes invites failure.
　　4. The attack on the freighter which was sunk was conducted in such a way as to indicate that again ranges were in error. If ranges had been more accurate, a more favorable firing position would have been obtained. It appears that the firing mechanism of the first torpedo failed to arm due to the short run. The subsequent tactics used in evading the escort were well conducted.
　　5. To get in an attack on such a valuable target as an aircraft carrier calls for the greatest degree of aggressiveness. The fact that the *Ryujo* was not sighted until at a range of 11,000 yards bears out the Commanding Officer's statement that a more alert watch could have been kept.
　　6. It is regretted that so much difficulty was encountered in the operation of the SJ Radar. This has greatly handicapped the *Wahoo* since valuable training in the use of this equipment has been missed. During the current refit period, Radar personnel should receive special instruction in its use. A training motor will be installed if available.

7. The excellent material condition of the *Wahoo* upon return from patrol is commendable. The following damage is considered as done the enemy:

SUNK

1 Freighter.............6,500 tons.

CONFIDENTIAL

From: The Commander Submarine Division ONE HUNDRED TWO.
To: The Commander Submarine Force, Pacific Fleet.

Subject: U.S.S. *Wahoo* (SS-238); First War Patrol, Comments on

1. The *Wahoo* spent 34 days in or in vicinity of the assigned area. During this period five worthwhile torpedo targets were contacted. Two of the targets were extremely valuable vessels, the *Ryujo* and *Chiyoda*. Of the five, one 6,500 ton freighter was sunk. More targets probably would have been sighted had Piaanu Pass been kept under closer observation then shown in the track chart.

2. Japanese tankers and other fast auxiliaries have been observed to be carrying depth charges which they apparently drop either to embarrass the attacking submarine or to countermine approaching torpedoes. This might have been the tactics employed by the small tanker attacked on the morning of September 6th and is offered as an explanation of the two delayed explosions which were heard one minute after torpedoes were estimated to have crossed the track.

3. It is very unfortunate that attack positions were not attained on either the *Chiyoda* nor *Ryujo*. In the case of the former, luck alone was the factor that saved her from certain destruction or severe damage. In the case of the latter, the situation upon sighting indicated that attack position, if attainable, could be gained only by the most aggressive kind of approach. It is regrettable that the need for such immediate action was not recognized.

4. S.J. Radar. It is regretted that so much difficulty was experienced with the S.J. Radar. Had the radar been kept in operation condition it might well have made the desired contact with *Ryujo* at noon on October 5th. No effort should be spared to keep this valuable instrument in perfect operating condition throughout a patrol. In regard to the tuning of the system it is suggested that the proximity of any high land in the patrol area at night will provide the "willing" target required for this purpose. The urgent need for trained radar maintenance men in submarines is again noted.

5. While the patrol was not outstanding in either results or aggressiveness it is certain that the resultant seasoning of the officers and crew has prepared them to most any situation on future patrols with resolution and effectiveness.

U.S.S. *Wahoo* (SS-238)

6. MATERIAL. *Wahoo* will be refitted by *Sperry* assisted as necessary by the Navy Yard.

It is assumed that steps have been taken to prevent a repetition of the gross carelessness which placed tube number 1 out of commission.

Serial 01249

Care of Fleet Post Office,
San Francisco, California,
1 November 1942

COMSUBPAC PATROL REPORT NO. 83
U.S.S. WAHOO - FIRST WAR PATROL
CONFIDENTIAL

From: The Commander Submarine Force, Pacific Fleet.
To: Submarine Force, Pacific Fleet.

Subject: U.S.S. *Wahoo* (SS-238) – Report of First War Patrol.

Enclosure: (A) CSS-10 Ltr.File FC-10/.16-3(043) of 26 Oct. 1942
(B) CSD-102 Ltr. FB5-102/A16-3(1) Serial 016 of October 20, 1942.
(C) Subject Patrol Report.

1. The Commander Submarine Force, Pacific Fleet, agrees with the comments of Commander Submarine Division 102 and Commander Submarine Squadron Ten that more time on this patrol should have been spent in close proximity to Piaanu Pass.

2. A successful attack on the *Ryujo* would have had far reaching results and it is unfortunate that the *Wahoo* failed to press home an attack in this instance. Opportunities to attack an enemy carrier are few and must be exploited to the limit with due acceptance of the hazards involved. The Commanding Officer realized his mistake in the instance, as noted in paragraph 17 of the report. The lesson learned by this experience should impress on all Submarine Commanding Officers the necessity for continuous alertness, and quick, positive action immediately when a contact is made.

3. Firing of four single torpedoes in the attack on September 20th undoubtedly resulted in the needless expenditure of torpedoes over what would have been required to obtain the same results had a torpedo spread been used in the initial attack.

4. The *Wahoo* is credited with having inflicted the following damage on the enemy:

U.S.S. *Wahoo* (SS-238)

SUNK
1 Freighter   6,441 tons

R.H. ENGLISH.

DISTRIBUTION
(35CM-42)
List III: SS
Special
 P1(5), EN3(5), Z1(5),
 Comsublant (2)
 Comsubsowespac (2)

/s/E.R. Swinburne
E. R. SWINBURNE,
Flag Secretary

Patrol Two, 8 Nov 1942 – 26 Dec 1942

Subject: U.S.S. WAHOO - REPORT OF SECOND WAR PATROL. PERIOD FROM NOVEMBER 8, 1942 TO DECEMBER 26, 1942.

Area: Dog (East).

Operation Order: ComSubPac SECRET Dispatch 041947 of November 1942 and ComTaskFor 42 SECRET dispatch 150805 of November 1942.

PROLOGUE: Arrived Pearl Harbor on October 17, 1942, from first war patrol. Commenced refit on October 18 with U.S.S. *Sperry* repair forces. Shifted to Submarine Base, Pearl on October 22, to complete refit, which was completed on November 2. Three day training period and readiness for sea on November 8. Installed 4-inch gun and two 20-mm. guns.

1. NARRATIVE:

November 8—09:00 (VW) Underway for patrol in company with small escort vessel *P-28*. Made trim dive, received indoctrinational depth charge, made structural test firings of 4-inch gun, and fired 10 rounds of target ammunition for training. Sighted numerous planes and ships during the day. The escort returned to port at dark.

November 9— Sighted U.S. Navy patrol planes at 07:00, 07:10, and 12:50, all times Xray.

November 14— Passed to command of ComSoPac at zero hours zed at Lat. 7°-50' N; Long. 176°-15' E.

November 16— Having run submerged during part of 14th and all of 15th daylight periods in passing Mili, decided to run on the surface; Mili, Jaluit and Makin all being about 120 miles distant. At 10:20(M) contacted airplane at 6 miles on radar and submerged.

November 20— Arrived in patrol area Dog (East) as directed by ComTaskFor 150805 of November. Sea condition 6, wind force 6, visibility low, and rains frequent. Continued submerged patrol.

U.S.S. *Wahoo* (SS-238)

November 22— Sighted Bougainville Island to southwest at a distance of about 75 miles. Sea and wind moderating.
November 23—17:11 (K) O.O.D. sighted an object believed to be a periscope. It was in sight for but a few seconds, and no further evidence was noted which would indicate the presence of an enemy. Bougainville and Buka Islands were in sight at this time.
November 30— At 20:30(K) in Lat. 4°-55' [S]; Long. 154°-49'E sighted the smoke of a ship bearing 150°T. Distance estimated at 8,000 yards. Changed course to head for the smoke. The night was quite dark; sky partially overcast and threatening thunderstorms. Brilliant flashes of lightning at irregular intervals illuminated the sea and horizon on all bearings. The smoke and the target were not visible except during these flashes. At 20:40 a brilliant lightning flash revealed the source of the smoke: A high hull, low superstructure vessel of considerable size, giving the appearance of a lightly burdened freighter or transport; angle on the bow about 10° starboard, range about 6,000 yards. Neither sound nor radar were able to pick up the target. A destroyer escort was on station on the port bow of the target. Dived. At 20:43 sound operator reported echo ranging, long scale, at true bearing of 070°, 280° relative. Started swinging left. At 20:46 the second sound operator reported echo ranging on true bearing of 169°, and shortly thereafter gave a propeller count of 120 RPM on that bearing. As this was apparently the target group sighted we commenced swinging right. Commenced sound tracking. Great difficult was had in picking up the target or its escort by periscope, due to the necessity of being trained on the proper bearing at the instant of a lightning flash. Sound bearings proved inadequate for this until at 20:56 a flash revealed a destroyer bearing 216°T, angle on the bow 90° starboard, range estimated at 3000 yards. As gyro angles were about 50° right and range indeterminate, did not fire. Swung right for a straight shot on a large track, but could not swing fast enough to even get a reasonable shot. At 21:00 all echo ranging stopped. This was essentially a sound approach. Attack was essentially a sound approach. Attack position was lost by the time the first periscope information was obtained. Radar was not used because it failed originally to pick up target, and tests off Pearl Harbor showed that even at short ranges the entire conning tower and bridge structure must be out of the water to obtain a contact. The approach was unsuccessful partly due to inaccurate, inadequate, and confused sound information and partly due to the failure to appreciate the true nature of the approach until too late, clinging to the hope that lightning flashes would provide data for a more accurate approach.
December 2— At 00:28(K) while 18 miles east of Cape Henpan, sound picked up propellers bearing 245°T, which bearing changed progressively to 180°T in 4 minutes. Moon was shining brightly and visibility was sufficiently good to see Cape Henpan, and nothing could be sighted. Propeller beat veri-

fied by several operators at 130 RPM. Sound must have been made by some submerged object very close aboard, probably a fish.

December 7— Having patrolled the Buka-Kilinailau for seventeen (17) days with but one contact decided to move eastward and patrol the direct route between Truk and the Shortlands for a few days. At 20:36(K) in Lat. 5°-20' S; Long. 155°-55' E, picked up propellers on sound. Propellers started suddenly, worked up to about 120 RPM and faded slowly. There was a high background noise on that bearing for about 5 minutes, which faded gradually. It sounded as it we had flushed a stationary submarine, which dove on contact.

December 8— At 02:20(K) in Lat. 5°-20' S; Long. 156°-15' E, sound picked up echo ranging to northward. At 02:30 radar reported contact at 062°T at a range of 18,000 yards, which contact was then lost. Came to normal approach course. At 02:37 sound and radar contacted target and commenced tracking, radar data being intermittent. At 02:45 sighted target bearing 082°T, range 14,000 by radar, angle on the bow about 80° starboard. Target was a large tanker, loaded, and headed in the general direction of the Shortlands, zig-zagging. Echo ranging was heard continuously from the target's general direction. Target speed computed to be 13 knots. At 03:05 range had closed to 6,000 yards on a track and true bearing of 145° starboard when radar contacted the escort astern of tanker. At 03:07 echo ranging stopped. The approach being over, submerged to 40 feet and tracked by radar and sound. Kept radar contact on AO but lost it on escort at this depth. In analyzing this approach it is apparent that it was over at the time the target was sighted, a fact which was not realized for twenty minutes thereafter. The performance of the radar and sound were gratifying. This being the important target we had moved east to get, we now headed west to return to the passage between Buka and Kilinailau.

December 10— (Attack No. 1) At 14:57 while in Lat. 4°-56' S; Long. 154°-58' E, sighted heavy smoke bearing 293° T at distance of about 16 miles. For the first [time] in ten days sea conditions were ideal for attack, with roughened surface and many whitecaps. Shifted position to northward and got ahead of the formation. Sound conditions fair to poor. The source of the smoke turned out to be a convoy of three AKs[*], of tonnages approximating 8,500, 6,000, and 4,000, escorted by one Asashio class DD.[**] The ships were heavily loaded and headed for the Shortland area. The formation was zig-zagging by simultaneous ships movements with the DD patrolling a front about two (2) miles wide at a mean distance of about 1,000 yards

---

[*] AK – Military cargo ship. Japanese civilian merchant ships were also called "Marus," from the suffix *Maru* in merchant ships names. Military type designators were used for both naval and commercial ships. (See Appendix II for a more comprehensive list.)
[**] DD – Destroyer.

U.S.S. *Wahoo* (SS-238)

ahead of the leading AK. Originally we decided to attack the DD first, but although he passed us at a range of about 300 yards his maneuvers were too radical for a good shot. Picked out the largest AK as a target, swung to a large track to open the range, and at 16:27(K) fired a spread of four (4) torpedoes at a range of 700 yards on 120° starboard track, bow tubes. Three torpedoes hit, which was just as well because even then he took nearly two (2) hours to sink. The second target passed about 300 yards astern during this firing. Made a setup to fire the stern tubes at third AK, but DD got pretty close before we could fire. Started down. DD laid first depth charge pattern across our stern as we passed 120 feet. They were fairly close aboard. The main induction trunk flooded, the bridge speaker flooded, some lights were knocked out, a small circulation water line carried away in the pump room, and some odd nuts, bolts, paint, etc., flew around. He continued making passes and dropping depth charges. As we passed 250 feet and blew negative the gasket on the inboard vent carried away. The flood valve did not hold, and we went to 350 feet. By using negative vent stops and locking the flood valve closed by hand it remained dry the second time. Ran silent on reverse of convoy's course, maneuvering to avoid attacks. Depth charges dropped in varying numbers at following times: 16:30, 16:35, 16:36, 16:38, 16:45, 16:50, 16:51, 16:53, 16:56, 17:05, 17:17, 17:24, 17:26, 17:29, 17:30, 17:31, 17:32, 17:35, 17:43, and 17:45. Total charges dropped were about 40, some fairly close, but after 17:00 they began to fall further astern. At 17:26 came to periscope depth for observation. One AK was standing down the coast, one was just beyond the target picking up survivors, and the DD was patrolling the area dropping depth charges periodically. The target was on even keel, with both wells under water, about 2 feet of the bow and stern visible, and the high part of the bridge and stack visible. There were about ten (10) boats in the water. Observed target continue to settle until dark, and at 18:15 heard the bulkheads go. The watertight integrity of this ship must have been remarkable. Issued ration of ½ ounce rum to the crew. Surfaced after moonset at 20:30 and moved off to north. Target identified as being similar to *Syoei Maru*, which is listed in ONI-208-J as 5,624 gross tons and in *RECOGNITION OF JAPANESE MERCHANTMEN* dated February 12, 1942 as 8,748 gross tons. It was a pretty big ship. At 22:15 received SUBS 42 NR 73A concerning probable ships movements in our area. Assumed this to be the convoy already contacted and continued moving to northeast. At 23:40 received SUBS 42 NR 75A extending area. Started general movement in direction of new area. Decided to move in slowly to give the crew a chance to recover from effects of depth charging.

December 12— At 02:35 in Lat. 4°-29' N; Long. 156°-12' E, sound picked up a noise similar to echo ranging. Shortly thereafter a cargo ship was sighted and picked up by radar bearing 087° T, angle on the bow 90° port, range

U.S.S. *Wahoo* (SS-238)

10,000 yards. At 02:45 angle on the bow became 150° port. Radar track for 30 minutes gave a mean course of about 020° T, speed 13. Ship was plainly visible during this time, and we trailed on the quarter. As a course of 020° was heading for no known Japanese base, we expected a course change. Did not close range because of excellent visibility. Trailing was doing no good and a decision had to be reached prior to daylight. At 03:05 decided to get on his track to the Shortland area in case he was at a rendezvous with an escort and would proceed in that direction at daylight. He apparently continued to northeast, as that is the last we saw of him. Ship was of medium size with a single stack amidships, coal burner, and gave an appearance of being loaded. It is believed that the noise heard was a fathometer, and that he was unescorted. This noise was heard continuously after once picked up. Sound conditions were not good, no propellers were ever heard. The radar contact was fair to good, and once contact was established it gave fairly good information up to a range of about 12,000 yards. The bridge T.B.T.* was used as a check.

December 14— At 08:15(K) sighted hospital ship similar to *Manila Maru* in Lat. 6°-22' N; Long. 156°-13' E, heading for the Shortlands on course 190°. Ship was properly marked, was on a steady course at steady speed, was unescorted, and there were no aircraft in the air. This conformed to International Law. When identification was completed at a range of about 8,500 yards we broke off the approach and turned away. Sound conditions were bad. The ship passed us about 3,500 yards abeam and her propellers were never heard.

December 14— (Attack No. 2) At 13:21(K), sighted a submarine on the surface in Lat. 6°-30' S; Long. 156°-09' E, on course 015° departing the Shortland area. Range estimated at 3,000 yards, speed 12. We just had time to swing and shoot. Fired 6 minutes and 46 seconds after sighting. During the swing, submarine was positively identified as Japanese by the large flag and the designation *I-2* painted on the side of the conning tower. Firing range 800 yards, fired divergent spread of three torpedoes. First torpedo hit about 20 feet forward of conning tower 37 seconds after firing. Ship went down with personnel still on the bridge. Two and one half minutes after the torpedo explosion the submarine collapsed at deep depth with a noise considerably louder than the torpedo explosion. Apparently some of the W.T. doors had been shut. There was no counter-attack. At time of sighting the visibility was poor due to rain squalls, the sea in condition 3, and sound terrible. Even at 800 yards the targets propellers could not be heard. This attack was

---

* T.B.T. – Target Bearing Transmitter. A pair of binoculars mounted on a pelorus, used for taking bearings on targets from the bridge while a submarine is surfaced. These bearings are automatically transmitted to the conning tower by pressing a button on a handgrip.

brought to a successful end largely through the splendid coordination of four officers, whose performance was outstanding. They were:
Lieut. G.W. Grider - O.O.D. and Diving Officer
Lieut. R.H. O'Kane - A.A.O.
Lieut. R.W. Paine - T.D.C. Operator
Lt. Cmdr. D.W. Morton - A.A.O.

December 15-- Decided to let the area of the submarine sinking cool off so went over and looked into Kieta Harbor. There were no ships visible inside the port. While in that vicinity sighted the masts of a steamer at 15:35(K), ship hull down, headed in a generally northward direction. It had apparently come out of the Shortland area. Range was too great to determine the presence of an escort. Nothing was heard on sound. Masts were in sight for about 20 minutes. During the inspection of Kieta several tall towers resembling radio, directionfinder, or radar towers were noted, on the 460-meter peak of Bakawari Island.

December 17— At 02:05(K) while patrolling in Lat. 5°-45' S; Long 156°-13' E, picked up echo ranging. The moon had set and the night was clear and dark, with the sea a flat calm. Closed the sound by surface running and at 02:41 sighted a small ship believed to be a small destroyer or escort vessel. Range at sighting estimated to be 4,000 yards. Sound conditions were spotty, with the propeller sounds fading in and out—mostly out. Radar could not pick up the ship. At time of sighting we were about 20° abaft his beam, and while watching he zigged away. There were no ships in company. As a stern chase on the surface on an echo ranging anti-submarine vessel which is zigzagging has small merit, we broke off the approach.

December 19— Cleared area Dog (south) at 20:00(K) and directed by SUBS 42 - Serial 78 Affirm.

December 20— At 00:30(K) while about 30 miles East of Buka Island we picked up a plane by its motor noise. SD radar not manned at the time. Submerged for one hour. This was the first indication of aircraft activity we had encountered in the area. At 20:00(K) cleared area Dog (east) for Brisbane in accordance with SUBS 42 Serial 11 cast as modified by serial 78 affirm.

December 21— Sighted lights believed to be aircraft flares to westward of Buka Island at 00:10 and 00:22. At 01:38(K) contacted airplane on radar at 2 miles and submerged for one hour. At 16:50(K) sighted smoke bearing 048° T. Position Lat. 6°-20' S; Long. 154°-00' E. Closed on normal approach course until dark but never sighted any ships. At dark we were about ten miles into *Grouper*'s area. Broke off the approach and resumed assigned track. Ship apparently was en route from Rabaul to the Shortlands. Received SUBS serial 86 affirm at 20:15(K) requiring acknowledgement. Acknowledged at 03:30(K) on 23rd.

U.S.S. *Wahoo* (SS-238)

December 23— While running on surface in Lat. 12°-06' S; Long. 157°-02' E, picked up airplane on radar at 5 miles and then sighted it. Plane was flying high in a cloudy sky, proceeding in a southerly direction. He sighted us, turned and headed for as at a gliding angle. Fired emergency rocket and flare. He continued to close in and at a range of about two miles we submerged. Believed plane to be friendly, but we don't even let a friendly plane come close unless he gives a clue that he recognizes us. Stayed down for an hour and when we surfaced he was gone.

December 26— Passed Moreton Island light for surface into Brisbane at 03:30 (Love).

SUMMARY OF SUBMARINE ATTACKS

		Attack # 1	Attack # 2
1	Number of torpedoes fired:	4	4
2	Firing interval:	12"-9"-12"	9"-1
3	Point of aim:	M.O.T.	M.O.T.
4	Track angles:	122°S, 125°S, 127°S, 130°S	78°P, 80°P, 82°P
5	Depth setting:	First two: 15' Second two: 6'	All: 10'
6	Estimated draft:	23'	18'
7	Torpedo performance:	Normal	Normal
8	Estimated enemy speed:	11 kts.	12 kts.
9	Results of attack:	3 hits. Target sank.	1 hit. Target sank.
10	Evidence of sinking:	Visual.	Visual.
11	Spread employed:	Divergent: 0°, 6°R, 6°L, 6°L	Divergent: 0°, 4°L, 4°R
12	Estimated firing range:	700-800 yds.	850 yds.
13	Gyro angles:	355° 8° 359° 2°	359° 352° 357°

Detailed data required is listed in table above. There were eight (8) contacts and two (2) attacks. The two night contacts, on November 30th and December 12th, should have resulted in attacks, but we muffed the chances. Anyhow, we did learn something about night fighting, we hope.

2. WEATHER.

Pearl Harbor to Solomans.
Normal. Trade winds to Marshalls then variable light winds. Sea smooth, conditions 0 to 2. Usual tropical rain squalls at frequent intervals.

Off Solomans.
November 19-30 Ran into heavy sea with strong winds from westward, which lasted until the 22nd. Wind and sea then moderated and became variable in force and direction with short periods of calm during the shifts. Temperature remained in the middle 80s, and the humidity was low. Visibility varied, being excellent during the forenoon and limited by haze and rain squalls during afternoon and evening. In general, it was good submarine weather.
December 1-10 Sea became clam, varying from 0 to 1, with humidity increasing until it became uncomfortable. Winds were light and variable. Visibility spotty with sky usually overcast and local rain squalls prevalent.
December 10-20 Variable. Sea would change from calm through condition 3 and back and back to calm. Land mostly obscured by haze. Moon extremely bright with corresponding excellent night visibility. Humidity reasonable. Rain squalls were frequent.

En route to Brisbane, Australia.
Calm Seam with moderate swell. The normal Southeast Trades were encountered, the visibility remaining excellent day and night. Infrequent rain squalls were encountered.

3. TIDAL INFORMATION.

The currents were never predictable, but a general trend was sometimes noted. Off Buka, the currents were to the North or Northeast from Cape Henpan to the longitude of Kilinailau. Between latitudes 5°-00' S and 5°-30' S and longitudes 155°-30' E and 157°-00' E the currents were generally between Northeast and Southeast. In the area North of Bougainville Straits the currents were to South and Southeast, regardless of wind, sea, or tides. Drift varied from 0.4 to 1.0 knots.

4. NAVIGATIONAL AIDS.

Buka, Bougainville and Kilinailau Islands were in sight at various times and the peaks and tangents were useful in establishing an approximate position. Kilinailau was visible for 8 or 10 miles and showed up well at night. The Southern Reef is well covered with trees. The landmarks on Buka plotted fairly well, especially off Cape Henpan. A good fix was rarely obtained off

Bougainville. The peaks were usually obscured by haze and the tangents never seemed to be in the same place twice. The southeast coast of Bougainville Strait provided excellent landmarks for establishing position. A general land haze in the area prevented the full use of these landmarks. With the excellent night visibility prevalent no difficulty was experienced with ordinary celo-navigation. The navigation officer fixed our position once or twice each night using ordinary sextant.

5. DESCRIPTION OF ALL ENEMY WARSHIPS, MERCHANT VESSELS, PATROL VESSELS, AND SAMPANS SIGHTED INCLUDING POSITION, COURSE AND SPEED, AND TIME OF SIGHTING.

1. (a) 20:30(K) November 30.
   (b) One large AK with 1 or 2 DD escorts.
   (c) Lat 4°-55' S; Long. 154°-49' E.
   (d) Course 305°.
   (e) Speed 13 knots.
   (f) Empty cargo ship and escort from Shortlands to Rabaul.

2. (a) 02:20(K) December 8.
   (b) AO similar to KYOKUTO MARU, with escort.
   (c) Lat 5°-20' S; Long. 156°-15' E.
   (d) Course 160° - 220°.
   (e) Speed 18 knots.
   (f) Loaded tanker and escort from Empire to Shortlands.

3. (a) 16:30(K) December 10.
   (b) Convoy of 3 AKs escorted by one Asashio class DD.
   (c) Lat 4°-56' S; Long. 154°-58' E.
   (d) Course 090° - 135°.
   (e) Speed 11 knots.
   (f) Ships en route from Rabaul to Shortlands fully loaded. Sunk one AK of about 8,500 tons. Ship sunk was a one deck, split well, single stack freighter, with prominent kingpost forward and aft. Kingposts were of the goalpost type with mast in the center. Stack was also very prominent for height and lack of surrounding superstructure. Tentatively identified as similar to *Syoei Maru* and believed to be about 8,500 tons.

4. (a) 02:35(K) December 12.
   (b) Medium sized cargo ship probably unescorted.
   (c) Lat 4°-29' N; Long. 156°-12' E.
   (d) Course 345° - 030°.
   (e) Speed 13 knots.

(f) Picked up sound similar to echo ranging which is believed to have been made by fathometer. Ship apparently en route Empire from Shortlands, but seemed to be loaded.

5. (a) 08:15(K) December 14.
 (b) Hospital ship similar to *Manila Maru*.
 (c) Lat 6°-22' N; Long. 156°-13' E.
 (d) Course 190°.
 (e) Speed 12 knots.
 (f) Ship conformed to Geneva Convention and in accordance with directives of ComTaskFor 7 we did not attack.

6. (a) 13:21(K) December 14.
 (b) Japanese Submarine *I-2**
 (c) Lat 6°-30' S; Long. 150°-09' E.
 (d) Course 015°.
 (e) Speed 11 knots.
 (f) Fired three torpedoes and sunk Target. He was proceeding singly on the surface leaving the Shortlands.

7. (a) 15:35(K) December 15.
 (b) Medium sized steamship.
 (c) Lat 6°-00' N; Long. 156°-05' E.
 (d) Course - northerly.
 (e) Speed - moderate.
 (f) Sighted masts and stack of ship hull down.

8. (a) 02:05(K) December 17.
 (b) Small DD or escort vessel.
 (c) Lat 5°-45' S; Long. 156°-13' E.
 (d) Course - southerly.
 (e) Speed - moderate.
 (f) Proceeding singly in the direction of Shortlands.

6. DESCRIPTION OF ALL AIRCRAFT SIGHTED, INCLUDING TYPE, POSITION, COURSE, ALTITUDE AND TIME OF SIGHTING.

1. (a) Time and date - November 8.
 (b) Type - Various.

---

* This sinking was disallowed post-war. If *Wahoo* did sink a Japanese submarine on 14 December 1942, it was not *I-2*, which was sunk by U.S.S. *Saufley* (DD-465) in the Bismarck Sea on 7 April 1944.

(c) Position - Off Pearl Harbor.
(d) Course - Various.
(e) Altitude - Various.
(f) Remarks - Normal operating planes.

2. (a) November 9 at 07:00, 07:10 and 12:50.
   (b) U.S. Navy PBY
   (c) 200 - 250 miles SW of Pearl Harbor.
   (d) SW in AM - NE in PM.
   (e) 1000 - 2000 feet.
   (f) Routine Patrols.

3. (a) November 16 at 10:20(M).
   (b)
   (c) 150 miles SW of Mili Atoll.
   (d)
   (e)
   (f) Radar contact at 6 miles.

4. (a) 00:30(K) December 20.
   (b)
   (c) 30 miles East of BUKA.
   (d)
   (e)
   (f) Picked up by motor noise.

5. (a) 0138(K) December 21.
   (b)
   (c) 50 miles west of BUKA.
   (d)
   (e)
   (f) Radar contact.

6. (a) 1110(K) December 23.
   (b) Similar to British "Albermarle I" bombers.
   (c) Lat. 11°-50' S; Long. 157°-02' E.
   (d) Various.
   (e) 10,000 to 15,000 feet.
   (f) First picked up by radar at 4 miles. Moved out to five and a half miles, then came in. Fired one recognition signal without apparent effect.

U.S.S. *Wahoo* (SS-238)

7. SUMMARY OF S/M ATTACKS.

Listed under paragraph 1 NARRATIVE.

8. ENEMY A/S MEASURES.

(a) Escort for AK sighted on November 20th apparently used echo ranging from about 20:45 to 21:00, was silent then started echo ranging again about 21:45. Echo ranging on 17 KCs, with pings at 8-second intervals. Passed about 3,000 yards abeam without detection. Zig-zagging.

(b) Escort for AO sighted on December 8th was echo-ranging continuously until we reached a point 6,000 yards on his quarter, at which time echo ranging ceased. Zig-zagging.

(c) Convoy encountered on December 10th was zig-zagging by simultaneous ships movements in obedience to flag hoists on escorting DD. Escort patrolled area across the front of the formation at high speed. No echo ranging was heard. Depth charge attacks after sinking of the one AK lasted for one hour 15 minutes, and DD probably expended all its depth charges. After the first four or five attacks had been delivered in rapid succession around the firing point, DD periodically stopped and listened for contact. Apparently our position was undetected once we cleared firing point.

(d) Cargo ship contacted on December 12 was zig-zagging, and was making a sound similar to echo ranging which is believed to have been a fathometer. No escort was detected.

(e) The small DD or escort vessel encountered on December 17th was echo ranging and zig-zagging. As he has [was?] headed in the general direction of the Shortlands it is presumed that he had released a convoy to the northward and was proceeding to port.

9. DESCRIPTION OF ENEMY MINE SWEEPING OPERATIONS.

No minecraft or mining operations were noted.

10. MAJOR DEFECTS EXPERIENCED.

The gasket on the negative tank inboard vent carried away when the vent was opened under air pressure of 120 lb.in^2. A retainer should be installed on [the] gasket similar to that on the flood valve.

11. COMMUNICATION.

Radio reception was very good and was compete. Bells were copied on 44.8 KCs and on the 5-megacycle band. The other frequencies were not as good and were seldom used. Attempted to use the underwater loop for reception submerged on December 1. Keel depth 56 feet; depth of loop 19 feet; distance to transmitting station 1,200 miles; frequency 44.6 KCs. Faint signals were heard, but were unable to copy through high noise level. Used the loop for copying on surface and when running at 40 feet; signals were generally readable down to a depth of 55 feet.

Last serial received 72. A 86. C 14. D 1. (Dec. 22nd)
Last serial sent 250305.

12. SOUND CONDITIONS AND DENSITY LAYERS.

Sound conditions varied from very poor to excellent, with the conditions getting progressively worse near land. The two extreme conditions were encountered on December 8th and 14th. On the 8th propellers were heard at 15,000 yards, the location being 65 miles from land. On the 14th propellers could not be heard at 800 yards, the location being 5 miles off the channel of Bougainville Straits. The water was laden with vegetable matter in suspension, the quantity increasing as the shoreline was approached. This resulted in a large number of fish, which were seen and heard almost constantly. Fish noises caused the sound operators considerable trouble until they learned to recognize the variations. Besides the usual clicks, wheezes, and whistles previously encountered, we frequently picked up a noise similar to a reciprocating engine with a loose bearing making from 120 to 140 RPM. This turned out to be from whales. All observed temperature gradients were zero with a water temperature of 85°. Density layers seemed to be present at various times, but were never highly pronounced. A slight increase in speed or change in variable water was sufficient to change depth through all such layers encountered.

13. HEALTH AND HABITABILITY.

The health of the crew was excellent throughout this patrol. Only six persons received treatment for colds, none of which passed the "sniffling" stage. The only serious illness was one case of Cellulitis, left ankle, which was treated with hot MGSO4 dressings and a short course of Sulfathiazole. Habitability was excellent during the entire patrol due to good functioning of the air conditioning plant. When the air conditioning units were shut down during the depth charge attack a marked discomfort was noted throughout the boat within a few minutes, although some sweating would doubtless have been noticeable in any case. The average submerged temperature was 85° F.

14. MILES STEAMED EN ROUTE TO AND FROM STATION.

Miles steamed to station, 2,987.
Time en route, 264.5 hours.
Average speed en route, 11.3 knots.
Miles steamed from station, 1,515.
Time en route, 127 hours.
Average speed en route, 11.9 knots.

15. FUEL EXPENDED.

Fuel used en route to station 29,912 gallons.
Fuel rate of consumption 10.0 gal. 1 mile.
Fuel expended on station 9,430 gallons.
Fuel expended en route Brisbane 26,428 gallons.
Fuel rate of consumption 17.4 gal. 1 mile.

16. FACTORS OF ENDURANCE REMAINING.

(a) Torpedoes - 17
(b) Fuel - 26,050
(c) Provisions - 30 days.
(d) Fresh water - Unlimited.
(e) Personnel - 15 days.

17. TERMINATION.

Patrol was ended by the provisions of the operation order. No factor of endurance was reached.

18. REMARKS.

Our weakness in night fighting was clearly brought out by this patrol. Although we had sunk one ship on our previous patrol at night, we missed two on this patrol through lack of perspective and just plain confusion. The experience gained should make us more adept at this type of attack on future patrols. From the contacts we made it is believed that much of the shipping into the Shortland area is coming direct from the Empire. The general track seemed to be between course 000° and 015° with a focal point at about Lat. 6°-30' S; Long. 156°-10' E.

(Endorsement to second war patrol report)

TF42/A16-3   TASK FORCE FORTY-TWO
Serial   01249   Care of Fleet Post Office,
                  San Francisco, California

December 28, 1942

S-E-C-R-E-T

From: The Commander Task Force Forty-Two.
To: The Commander in Chief, United States Fleet.
Via: The Commander Submarine Force, Pacific Fleet.

Subject: U.S.S. *Wahoo* (SS-238), Second War Patrol, Comments on.

Enclosure: (A) Copy of Subject Patrol Report.

1. Enclosure (A) is forwarded herewith.

2. The *Wahoo* completed her second war patrol on December 26, 1942, having spent 46 days at sea and 29 days in the assigned area.

3. Eight contacts were made, of which only two were developed into attacks, both of which resulted in sinkings. However, it is believed that at least three of the other contacts should have been developed into attacks; namely on 30 November on the freighter, on 8 December on the large tanker, and again on December 12, speed should have been used to close this apparently unescorted vessel. It is noted that the radar functioned exceptionally well and it appears that this information was not used to best advantage to develop these contacts.

4. Sound conditions varied from poor to good; generally, however, they were poor.

5. The *Wahoo* returned in excellent material condition. The current refit will be accomplished by the SPERRY (AS-12).

U.S.S. *Wahoo* (SS-238)

6. The *Wahoo* is congratulated on sinking:

   1 freighter of *Syoei Maru* Class – 5,644 tons.
   1 submarine (*I-2*) – 1,955 tons.

                              JAMES FIFE, jr.

DISTRIBUTION
VCNO, Cinclant, Cincpac,
Comsowespac, Comsublant,
CSS 8 & 10, CSD 102, *Wahoo* File,
Each SS TF42 (not to be taken to sea, BURN),
Patrol Summary File, War Diary.

Editor's Note:

This was Kennedy's last patrol in *Wahoo*. Several officers expressed disappointment in the way contacts were not followed up, and requested transfer. In his endorsement to the patrol report, Admiral Fife also suggested that a more aggressive pursuit would have been appropriate.

In the end, rather than transfer the unhappy officers, Fife relieved Kennedy and gave *Wahoo* to Lieutenant Commander Dudley W. "Mush" Morton, who was mentioned by Kennedy in the attack on what was believed to be the Japanese submarine *I-2*. Morton was aboard on this patrol for a PCO (Prospective Commanding Officer) cruise, and would normally have gone on to command of another boat following the patrol. (A number of sources, including, rather embarrassingly, *The Dictionary of American Naval Fighting Ships*, incorrectly indicate that Morton was Kennedy's Executive Officer. He was not.)

Morton's Executive Officer was Lieutenant Richard H. O'Kane, who had also been Kennedy's Executive Officer, and had previously served as temporary XO under Stephen Barchet in *Argonaut*, during that boat's voyage to Mare Island for overhaul following her first war patrol. Together, Morton and O'Kane made up an exceptional team. Morton would become the number three commander in terms of enemy ships sunk (17 credited, 19 confirmed post-war). O'Kane, after leaving *Wahoo* for *Tang*, would go on to take the number one position, with 31 ships credited during the war and 24 confirmed post-war.

Patrol Three, 16 Jan 1943 – 7 Feb 1943

Subject: U.S.S. *Wahoo* – Third War Patrol – Report of.

PROLOGUE TO

    Arrived Brisbane, Queensland, Australia, December 26, 1942 after Second War Patrol and moored alongside USS *Sperry*. On December 27, 1942, commenced refit by USS *Sperry*, relief crew and ship's force. Refit consisted mostly of routine items plus a few minor repairs.
    On December 31, 1942 Lieut. Comdr. M.G. Kennedy was relieved as Commanding Officer by Lieut. Comdr. D. W. Morton.
    Ship ready for sea on January 16, 1943.

1. Narrative:

January 16th:
    09:00(L) Departed Brisbane, Queensland, Australia.
    10:30(L) Commenced sound listening tests in Moreton Bay.
    15:00(L) Completed sound tests.
    17:00(L) Transferred pilot and fell in company with our escort, USS *Patterson*.
    19:45(L) Made trim dive.
    20:30(L) Commenced night surface runs on our escort.
    23:06(L) Completed runs. Set course for area at two-engine speed (80-90). Still in company with our escort.

January 17th:
    08:07(L) Dived. Commenced DD-SS run for USS *Patterson*.
    11:00(L) Made deep dive; no leaks.

13:35(L) Dived. Commenced torpedo practice approaches on our escort.

14:45(L) Upon surfacing and while starting #2 engine for propulsion, flooded same, and put it out of commission (See Derangement Report – Page 17).

17:28(L) Completed runs. Escort departed. Set course for area at two engine speed (80-90).

January 18th:

13:15(L) [03:15?] Exchanged recognition signals with USS *Grampus*. ComTaskForce-Two had advised us both that we would pass during the night. 10:30(L) #2 engine back in commission. 14:00L Set clocks back –10 zone time. Conducted drills submerged and made frequent battle surfaces firing both 20-mm guns and 4" gun while en route to area.

January 19th:

22:00(K) Speeded up to three-engine speed (80-90) in order to make the passage in Vitiaz Straits during daylight. This will also, give us an additional day to cover Wewak and still arrive in area as directed. The additional fuel thus used is considered to be wisely expended.

January 21st:

18:20(K) Dived on SD radar contact. Upon reaching 70 feet stern planes jammed on hard rise causing us to broach at 30° up angle. Fortunately SD contact was false, the pip being an internal disturbance.

PROLOGUE TO WEWAK

Our Operation Order routed us through the vicinity of Wewak, a more or less undetermined spot located in whole degrees of latitude and longitude as 4° S and 144° E. Air reconnaissance had reported considerable shipping there, and dispatches received en route indicated continued use of this area by the enemy. The position of Wewak Harbor was determined as behind Kairiru and Nushu Islands on the Northwest Coast of New Guinea through the interest of D.C. Keeter, MM1C, U.S. Navy, who had purchased an Australian "two-bit" school atlas of the area.

Study of the harbor on our small-scale chart immediately showed deep water and unmistakable landmarks, with tempting possibilities for penetration and escape. By making an accurate tracing slide, and using a camera and signal light as a projector, a large scale chart was constructed of the whole harbor. All available information was transferred from sailing directions to this chart. With everything in readiness adjusted speed to arrive off Kairiru Island prior to dawn.

U.S.S. *Wahoo* (SS-238)

(All times K)

January 24th:
 03:30 Dived two and a half miles North of Kairiru Island and proceeded around western end to investigate Victoria Bay. As Dawn was breaking, sighted small tug with barge alongside and a few moments later two *Chidori* class torpedo boats. As this patrol was underway, maneuvered to avoid, then came back for a better look into this mile deep bay. There was no other shipping.

 Went around [the] southwestern tip of Kairiru Island to observe the strait between this and Nushu Island, a foul weather anchorage. Kept position out in, noting the set and drift, and light patches of water marking shallows. The water in general was a dirty yellowish green. With these in mind planned appropriate exit.

 Saw what appeared to be tripod masts on the eastern end of Karsau Island, but either a patrol boat or tug in Kairiru Strait prevented further observation at this time. As the masts could well have been those of a ship behind Karsau Island, proceeded west hoping to round Unei Island, connected to Karsau by a reef, and observe from between these islands and the mainland. However a reef with the seas breaking over it extended far west of Unei frustrated this plan. Went back between Karsau and Kairiru Islands hoping to see further around the eastern end. The masts were not sighted again, but a photograph taken at their observation may yet disclose their presence.

 At 13:18 an object was sighted in the bight* of Nushu Island, about five miles farther into the harbor, much resembling the bridge-structure of a ship. Commenced approach at three knots. As the range closed the aspect of the target changed from that of a tender with several small ships alongside to that of a destroyer with *RO* class submarines nested, the latter identified by the canvas hatch hoods and awnings shown in ONI 14. The meager observations permissible were insufficient for positive identification and the objects alongside may have been the tug and barge first sighted at dawn in Victoria Bay.

 It was our intention to fire high-speed shots from about 3,000 yards, which would permit us to remain in deep water and facilitate an exit. However, on the next observation, when the generated range was 3,750, our target, a *Pubuki* class destroyer was underway. Angle on the bow 10 port, range 3,100. Nothing else was in sight. Maneuvered for a stern tube shot, but on next observation target had zigged left giving us a bow tube set up.

 At 14:41 fired spread of three torpedoes on 110° starboard track, range 1,800 yards, using target speed fifteen since there had been insufficient time

---

* The original report has "hight" here. Bight appears more likely to be the intended meaning than height, given the context.

to determine speed by tracking. Observed torpedoes going aft as sound indicated 18 knots, so fired another fish with enemy speed 20. Destroyer avoided by turning away, then circled to the right and headed for us. Watched him come and kept bow pointed at him. Delayed firing our fifth torpedo until the destroyer had closed to about 1,200 yards, angle on the bow 10° starboard. Then to insure maximum likelihood of hitting with our last torpedo on the forward tubes, withheld fire until range was about 800 yards. This last one, fired at 14:49, clipped him amidships in twenty-five seconds and broke his back. The explosion was terrific!

The topside was covered with Japs on turret tops and in the rigging. Over 100 members of the crew must have been acting as lookouts.

We took several pictures, and as her bow was settling fast we went to 150 feet and commenced the nine-mile trip out of Wewak. Heard her boilers go in between the noise of continuous shelling from somewhere plus a couple of aerial bombs. They were evidently trying to make us lie on the bottom until their patrol boats could return.

No difficulty was experienced in piloting without observation out of Wewak using sound bearings of beach noises on reefs and beach-heads. With the aid of a one-knot set we surfaced at 19:30 well clear of Kairiru and Valif Islands. Cleared area on four engines for 30 minutes on course 000° T. Huge fires were visible in Wewak Harbor. We wondered if they had purposely created these fires to silhouette us in case we tried to escape out of the harbor. Slowed to one engine speed (80-90) at 20:00.

22:30 As the enemy convoy route from Palau to Wewak was known to pass between Wuvulu and Aua Islands commenced search by criss-crossing base course at 30 degrees on two hour legs. 23:45 Sent report of Wewak engagement to ComTaskForce Forty-Two.

January 25th: (All times K).

05:30 Passed between Aua and Wuvulu Islands. Changed base course for Palau and went to two engine speed (80-90) continuing the criss-cross search for enemy shipping.

08:30 Fired Tommy gun across bow of small fishing boat and brought him alongside. Neither our Chamoro or Filipino mess boy could converse with the six Malayans in the boat, but by sign language we learned that they were originally nine in number, three having died. One of the remaining six was apparently blind, a second quite sick, and a third obviously suffering from scurvy. Gave them food and water as they had none and then continued our search for the enemy.

10:00 In accordance with Operation Order, shifted from Task Force Forty-Two to SubPacFor without dispatch. Commenced guarding SubPac radio schedules.

U.S.S. *Wahoo* (SS-238)

16:45 Dived for a half-hour and held various drills. While submerged passed under the equator.

January 26th: (All times K).

07:57 Sighted smoke on the horizon, swung ship towards and commenced surface tracking. Adjusted course and speed to get ahead of the enemy. After three quarters of an hour and when we had obtained a favorable position with masts of two ships just coming over the horizon, dived and commenced submerged approach.

The two freighters were tracked at 10 knots on a steady course of 095° T., which was somewhat puzzling as it led neither to nor from a known port. During the approach determined that the best firing position would be 1,300 yards on beam of leading ship. This would permit firing with about 15° right gyro angle on approximately a 105° track on the leading ship, and with about 30° left gyro angle and 60° track on the second ship 1,000 yards astern in column. However at 10:30 found we were too close to the track for this two ship shot so reversed course to the right and obtained an identical set-up for a stern tube shot. At 10:41 fired two torpedoes at the leading ship and seventeen seconds later two at the second freighter. The first two torpedoes hit their points of aim in bow and stern. There was insufficient time allowed for the gyro setting angle indicator and regulator to catch up with the new set-up cranked into the TDC for the third shot. This torpedo passed ahead of the second target. The fourth torpedo hit him. Swung left to bring bow tubes to bear in case these ships did not sink.

At 10:45 took sweep around to keep the set-up at hand and observed three ships close about us. Our first target was listed badly to starboard and sinking by the stern, our second was heading directly for us, but at slow speed, and the third was a huge transport which had evidently been beyond and behind our second target.

At 10:47, when the transport presented a 90° starboard angle on the bow at 1,800 yards range fired spread of three torpedoes from forward tubes. The second and third torpedoes hit and stopped him. We then turned our attention to the second target, which was last observed heading for us. He was still coming, yawing somewhat, and quite close. Fired two bow torpedoes down his throat to stop him, and as a defensive move. The second torpedo hit, but he kept coming and forced us to turn hard left, duck and go at full speed to avoid.

There followed so many explosions that it was impossible to tell just what was taking place. Eight minutes later came back to periscope depth, after reaching 80 feet, to observe that our first target had sunk, our second target still going, but slowly and with evident steering trouble, and the transport stopped but still afloat. Headed for transport and maneuvered for a killer shot.

U.S.S. *Wahoo* (SS-238)

At 11:33 fired a bow torpedo at 1,000 yards range, 85° port track, target stopped. The torpedo wake passed directly under the middle of the ship, but the torpedo failed to explode. The transport was firing continuously at the periscope and torpedo wake with deck guns and rifles.

At 11:35 fired a second torpedo with the same set-up except that the transport had moved ahead a little and turned towards presenting a 65° angle on the bow. The torpedo wake headed right for his stack. The explosion blew her midships section higher than a kite. Troops commenced jumping over the side like ants off a hot plate. Her stern went up and she headed for the bottom. Took several pictures.

At 11:36 swung ship and headed for the cripple, our second target, which was now going away on course 085°. Tracked her at six knots, but could not close her as our battery was getting low.

At 11:55 sighted tops of fourth ship to the right of the cripple. Her thick masts in line had the appearance of a light cruiser's tops. Kept heading for these ships hoping that the last one sighted would attempt to pick up survivors of the transport. When the range was about 10,000 yards, however, she turned right and joined the cripple, her masts, bridge structure and engines aft identifying her as a tanker. Decided to let these two ships get over the horizon while we surfaced to charge batteries and destroy the estimated twenty troop boats now in the water. These boats were of many types, scows, motor launches, cabin cruisers and nondescript varieties.

At 11:35 made battle surface and manned all guns. Fired 4" gun at largest scow loaded with troops. Although all troops in this boat apparently jumped in the water our fire was returned by small caliber machine guns. We then opened fire with everything we had. Then set course 085° at flank speed to overtake the cripple and tanker.

At 15:30 sighted smoke of the fleeing ships a point on the port bow. Changed course to intercept. Closed until the mast tops of both ships were in sight and tracked them on course 350°. They had changed course about 90° to the left, apparently to give us the slip. Maneuvered by mooring board to get ahead undetected, but kept mastheads in sight continuously by utilizing No. 1 periscope and locating look-out on top of periscope shears. At 17:21, one half hour before sunset, with the two ship's masts in line, dived and commenced submerged approach. Target zigs necessitated very high submerged speeds to close the range. Someone said the pitometer log indicated as much as 10 knots. Decided to attack tanker first, if opportunity permitted, as she was yet undamaged.

At 18:29, when it was too dark to take a periscope range, fired a spread of three bow torpedoes with generated range 2,300 yards, on a 110° port track. One good hit was observed and heard one minute, twenty-two seconds after firing. This apparently stopped him. Started swing for stern tube shot on the freighter but he had turned away.

U.S.S. *Wahoo* (SS-238)

Surfaced twelve minutes after firing and went after the freighter. Was surprised to see the tanker we had just hit still going and on the freighter's quarter. We were most fortunate to have a dark night with the moonrise not until 21:32, and to have targets that persisted in staying together. Our only handicap was having only four torpedoes left, and these in the stern tubes.

Made numerous approaches on the tanker first, as he was not firing at us. Even attempted backing in at full speed, but the ship would not answer her rudder quickly enough. After an hour and a half was able to diagnose their tactics. Closed in on tanker from directly astern, when they zigged to the right we held our course and speed. When they zigged back to the left we were on parallel course at 2,000 yards range. Converged a little on the tankers port beam, then twisted left with full rudder and power. He thus gave us a stern tube shot, range 1,850 yards on a 90° port track.

At 20:25 fired two torpedoes at tanker, the second hitting him just abaft of his midships, breaking his back. He went down in the middle almost instantly.

Immediately after firing changed course to head for the freighter and went ahead full. Passed the tanker at 1,250 yards by SJ radar, at which time he occupied full field in 7x50 binoculars. This fixes his length at about 500 feet. Only the bow section was afloat and its mast canted over when we left him astern.

At 20:36, eleven minutes after firing on the tanker, commenced approach on our last target. It was quite evident that this freighter had a good crew aboard. They did not miss an opportunity to upset our approach by zigs, and kept up incessant gunfire to keep us away. Much of this firing was at random, but at 20:43 they got our range, placed a shell directly in front of us, which ricocheted over our heads and forced us to dive.

Our "gun-club"* could take a lesson from their powder manufacturers. It was truly flashless, a glow about the intensity of a dimmed flash-light being the only indication that a projectile was on its way. It is somewhat disconcerting when a splash is the first indication you are being fired upon.

We tracked the freighter by sound until the noise of shell splashes let up then surfaced at 20:58, fifteen minutes after diving, and went after him. Two minutes later a large searchlight commenced sweeping sharp on our port bow, its rays seemingly just clearing our periscope shears. Assumed this was from a man-of-war and that the freighter would close it for protection. Our attack obviously had to be completed in a hurry. Headed for the searchlight beam and was most fortunate to have the freighter follow suit.

At 21:10, when the range was 2,900 yards by radar, twisted to the left for a straight stern shot, stopped and steadied. Three minutes later, with an-

---

* Gun Club – The Navy Bureau of Ordnance (BuOrd).

gle on the bow 135° port by radar tracking, fired our last two torpedoes without spread. They both hit, the explosions even jarring us on the bridge.

As the belated escort was now coming over the horizon, silhouetting the freighter in her searchlight, we headed away to the east and then five minutes later to the north. Fifteen minutes after firing the freighter sank leaving only the destroyer's searchlight sweeping a clear horizon. It had required four hits from three separate attacks to sink this ship.

At 21:30 set course 358° for Fais Island. At 23:45 sent dispatch to ComSubPac concerning new route and engagement.

Two men were injured by 20-mm explosion. The cause is covered in the Report of investigation and treatment in the Health and Habitability Report, included herewith.

January 27th: (All times K):

07:20 Sighted smoke over the horizon, commenced tracking and changed course to intercept. At 08:01, when masts of three ships were in sight, dived and continued approach. The mean course was plotted as 146° with the whole convoy zigging simultaneously thirty degrees either side of base course. At 08:30 the tops and stacks of two more freighters, and those of a tanker with engines aft were in sight.

It was our first intention to intercept one of the lagging freighters, which did not appear to be armed, but a zig placed the tanker closest to us. Surfaced with range about 12,000 yards and headed at full speed to cut him off. Trained gun sharp on starboard bow, then sent pointer and trainer below to stand by with rest of gun crew. The convoy sighted us in about 10 minutes, commenced smoking like a Winton*, and headed for a lone rain-squall. Only two of the larger freighters opened fire and their splashes were several thousand yards short. Their maneuver left the tanker trailing, just where we wanted him.

At 10:00, when we had closed to 7,500 yards, however, a single mast poked out from behind one of the smaller freighters. Almost immediately the upper works of a corvette or destroyer were in sight. Turned tail at full power to draw the escort as far as possible away from the convoy in case we were forced to dive, as this would greatly shorten the time he could remain behind to work us over.

Ordered contact report to be sent, but could not raise anyone.

Found that our engineers could add close to another knot to our speed when they knew we were being pursued. We actually made about 20 knots, opening the range to thirteen or fourteen thousand yards in the first twenty

---

* Winton— Also known as the GM-Winton, these V-16 diesels were one of the two primary types fitted in *Gato* class submarines. *Wahoo* was fitted with Fairbanks-Morse engines, which Morton evidently thought were superior—or, at least, smoked less.

minutes of the chase. In fact, he was smoking so profusely that we called him an "Antiquated Coal-burning Corvette." He was just lighting off more boilers evidently, for seventeen minutes later he changed our tune by boiling over the horizon, swinging left, and letting fly a broadside at estimated range of 7,000 yards. There was no doubt about his identity then, especially when the salvo whistled over our heads, the splashes landing about 500 yards directly ahead. Dived and as we passed periscope depth felt gun splashes directly over-head. Went to 300 feet and received six depth charges fifteen minutes later. They sounded loud, but did no damage.

Lost sound contact at 11:20. As the DD had some forty miles to catch up with his leading ships he evidently didn't stay around. We decided to catch our breath none-the-less, so stayed deep until 14:00 when we surfaced and commenced running again for Fais. At 20:58 sent contact report of convoy to ComSubPac.

January 28th: (All times K)

08:30 Sighted Fais Island fifteen miles ahead. Dived twenty minutes later on ten mile circle and closed the island at 4 knots. Took soundings with single signal at 10-minute intervals, and tried echo ranging on the reef. The soundings agreed closely with those on chart 5426. The echo ranging was unsuccessful due to bottom reverberations. There was no evidence of a sound listening post. The trading station is just as shown on the chart.

Proceeded around the southwestern end of the island one and a half miles from the beach and located the Phosphorite Works, warehouses and refinery on and inshore of the prominent point in the middle of the northwest side of the island.

Immediately made plans to shell these works that evening at moon-rise with our few remaining 4" rounds as the large refinery, warehouses, etc., offered a splendid target. This plan was frustrated by the arrival at 14:00 of an Inter-Island Steamer with efficient looking gun mounts forward and aft. She was similar to the sketch of the Q-boat, appearing in the *Gudgeon*'s Second Patrol Report, except that she was half again as long. Swung and moored to the buoy off the Refinery Point, where she would have made a nice target for one torpedo. Her tonnage was estimated at 2,000.

The phosphorite is evidently loaded from the crane, visible on the point, into lighters, which were observed moored well inshore, and thus transferred to the steamers. At 16:00 decided to leave well enough alone, so after taking several more photographs set course to clear northern end of Fais Island.

At 18:00 surfaced and went ahead on three main engines on prescribed route to Pearl Harbor, T.H.

U.S.S. *Wahoo* (SS-238)

February 7th: (All times V-W)
08:30 Arrived Pearl.

2. WEATHER:

Excellent weather was experienced throughout the patrol, with increasing seas on approaching Pearl Harbor, T.H.

3. TIDAL INFORMATION:

A one-knot southerly set was experienced in the approaches to and passage through Vitias Straits. This is contrary to all available information.

4. NAVAGATIONAL AIDS:

None.

5. ENEMY SHIPS SIGHTED:

Date	Time	Position	Course	Speed	Type
1/24/43	06:30K	Lat 3°- 23' S Long 143°-34' E	Maneuvering in harbor		2 *Chidori* Class torpedo boats.
1/24/43	13:18K	Lat 3°-23' S Long 143°-34' E	Anchored and maneuvering in harbor.	20	1 *Fubuki* Class destroyer.
1/26/43	08:45K	Lat 1°-55' N Long 139°-14' E	095° T	10	1 AK (*Dakar Maru*) 1 AK *(Arizona Maru)*
1/26/43	10:45K	Lat 1°-55' N Long 139°-14' E	095° T	10	1 AP *(Seiwa Maru)*
1/26/43	11:55k	Lat 1°-55' N Long 139°-05' E	350° T	10	1 AO *(Manzyu Maru)*
1/27/43	08:00K	Lat 4°-15' N Long 140°-05' E	146° T	9	5 AK 1 AO
1/28/43	14:00K	Lat 9°-45' N Long 140°-30' E	Underway and anchored.		1 Inter island similar to but longer than the Q-ship described in *Gudgeon*'s 2nd war patrol rpt.

6. AIRCRAFT SIGHTINGS:

None – one felt.

45

U.S.S. *Wahoo* (SS-238)

7. SUMMARY OF SUBMARINE ATTACKS:

Attack No.	1	2	3
Date	1/24/43	1/24/43	1/24/43
Location (Latitude)	3° 23' S	3° 23' S	3° 23' S
(Longitude)	143° 34' E	143° 34' E	143° 34' E
No. of torpedoes fired	3	1	2
Hits	0	0	1
Sunk (Tonnage)	—	—	1850
Damaged	—	—	—
Type of target	*Fubuki* Class Destroyer	*Fubuki* Class Destroyer	*Fubuki* Class Destroyer
Range of firing	1800	1900	800
Estimated draft of target	10'	10'	10'
Torpedo depth setting	2'	2'	2'
Bow or stern shot	Bow	Bow	Bow
Track angle	110° S	135° S	20° S
Gyro angles	358° 0° 2°	5°	15° 18°
Target speed used	15	20	20
Firing interval	11 sec, 12 sec	—	20 sec
Spread: amount and kind	2° divergent		None
Type of attack	Periscope	Periscope	Periscope

Attack No.	4	5	6
Date	1/26/43	1/26/43	1/26/43
Location (Latitude)	1° 55' N	1° 55' N	1° 55' N
(Longitude)	139° 14' E	139° 14' E	139° 14' E
No. of torpedoes fired	2	2	3
Hits	2	1	2
Sunk (Tonnage)	7,160	—	1850
Damaged	—	Yes	Yes
Type of target	*Dakar Maru* Cl freighter	*Arizona Maru* Cl freighter	*Seiwa Maru* Cl transport
Range of firing	1300	1550	1800
Estimated draft of target	20'	20'	20'
Torpedo depth setting	8'	8'	8'
Bow or stern shot	Stern	Stern	Bow
Track angle	110° S	70° S	80° S
Gyro angles	193° 195°	164°	344° 349° 345°
Target speed used	10	10	10
Firing interval	8 sec	11 sec	11 sec, 11 sec
Spread: amount and kind	Diverg. diff pts of aim	Diverg. diff pts of aim	2° divergent
Type of attack	Periscope	Periscope	Periscope

U.S.S. *Wahoo* (SS-238)

Attack No.	7	8	9
Date	1/26/43	1/26/43	1/26/43
Location (Latitude)	1° 55' N	1° 55' N	2° 34' N
(Longitude)	139° 14' E	139° 14' E	139° 25' E
No. of torpedoes fired	2	2	3
Hits	1	1 (+1 dud)	3
Sunk (Tonnage)	—	7,120	—
Damaged	Yes	—	Yes
Type of target	*Arizona Maru* cl freighter	*Seiwa Maru* cl transport	*Manzyu Maru* cl tanker
Range of firing	800	1100 & 900	2200
Estimated draft of target	20'	20'	20'
Torpedo depth setting	8'	8'	8'
Bow or stern shot	Bow	Bow	Bow
Track angle	20° S	50° S	110° S
Gyro angles	358° 349°	359° 359°	353° 351° 352°
Target speed used	9	9	9
Firing interval	10 sec	1 min, 26 sec	23 sec, 28 sec
Spread: amount and kind	None	None	1° divergent
Type of attack	Periscope	Periscope	Periscope

Attack No.	10	11
Date	1/26/43	1/26/43
Location (Latitude)	2° 37' N	2° 30' N
(Longitude)	139° 42' E	139° 44' E
No. of torpedoes fired	2	2
Hits	1	2
Sunk (Tonnage)	6,520	9,500
Damaged	—	
Type of target	*Manzyu Maru* cl tanker	*Arizona Maru* cl freighter
Range of firing	1,800	3,500
Estimated draft of target	20'	20'
Torpedo depth setting	2'	2'
Bow or stern shot	Stern	Stern
Track angle	100° S	145° S
Gyro angles	165° 165°	175° 175°
Target speed used	9	9
Firing interval	18 sec	15 sec
Spread: amount and kind	1° divergent	None
Type of attack	Night surf	Night surf

U.S.S. *Wahoo* (SS-238)

8. ENEMY A/S MEASURES:

The enemy apparently uses gunfire whenever possible as a nuisance factor to keep a submarine down. The fire of their merchantmen was in general inaccurate, but their destroyers should not be underestimated.

Aircraft bombs were evidently dropped as a nuisance factor also, in an attempt to make us lie on the bottom. The noise only is disturbing.

The depth charging consisted of a single pattern laid on the last known position, with destroyer speed about 35 knots.

9. MAJOR DEFECTS:

None.

10. COMMUNICATIONS:

Radio reception was very good and was complete. No attempt was made to use the underwater loop for submerged reception. Two messages were sent to ComSubPac from contact area. In neither case were we able to reach NPM and the message was given to an Australian station in each case.

Last serial received: SubPac Serial 14
Last message sent: *Wahoo* 052045 of February.

11. SOUND CONDITIONS AND DENSITY LAYERS:

Sound conditions were generally fair to poor. Propellers were picked up at ranges of 4,000 to 5,000 yards. At Wewak Beach noises were very distinct from each island and were used to check the DR position. While evading attack by destroyer January 27 in 4°-31' N; 140°-40' E, two density layers were encountered.

12. HEALTH AND HABITABILITY:

The health of the crew for this patrol can only be classified as "fair." In addition to the injuries there were a few cases of boils and numerous cases of colds. The latter can be attributed to the sudden changes of air conditions we were compelled to go through during the actions. The crew had been under considerable strain for about three days and their resistance had been definitely lowered.

The injuries were as follows:

One man received a severe laceration of the right forearm, which required seven stitches. Two men were injured by the misfire of the 20-mm gun. In one of these cases it was deemed necessary to amputate two toes of the right foot. Due to a shortage of surgical instruments a pair of sterilized side cutters were used to cut portions of the shattered bone. Because of the phalanges in the second toe being completely shattered it was not sutured closed but left open to allow free drainage. A generous amount of Sulfanilamide powder was used. The other man was wounded in the shoulder but, no lead or foreign body could be located. Three sutures were used in closing the laceration. This man was back to duty in three days with no complications.

Habitability was excellent.

13. MILES STEAMED:

Steamed 6,454 miles en route Brisbane to Pearl Harbor, plus approximately 100 miles at full power during attacks and counter-attacks. Total 6,554 miles.

14. FUEL OIL EXPENDED:

92,020 gallons; 14.1 gallons per mile.

15. ENDURANCE FACTORS:

Torpedoes – none. Other factors – indefinite.

16. PATROL ENDED:

By orders of ComSubPac, after expenditure of all torpedoes.

17. REMARKS:

(a) The fire control party of this ship was completely reorganized prior to and during this patrol. The Executive Officer, Lieutenant R.H. O'Kane is the co-approach officer. He made all observations through the periscope and fired all torpedoes. The Commanding Officer studies the various setups by the use of the IsWas[*] and analyzing the T.D.C. and does the conning. A third officer assists the Commanding Officer in analyzing the problem by studying the plot and the data sheets. On the surface the Executive Officer mans the T.B.T., and makes observations and does the firing. The Commanding Officer conns.

---

[*] A type of circular slide rule, which was used together with the "banjo" before the advent of the TDC to work out targeting solutions. It continued in use as a check on the TDC in many boats.

U.S.S. *Wahoo* (SS-238)

      This type of fire control party relieves the Commanding Officer of a lot of strain and it gives excellent training to all hands, especially the Executive Officer. It is recommended that other ships give it consideration and thought.

(b) The conduct and discipline of the officers and men of this ship while under fire were superb. They enjoyed nothing better than a good fight. I commend them all for a job well done, especially Lieutenant R.H. O'Kane, the Executive Officer, who is cool and deliberate under fire. O'Kane is the fightingest naval officer I have ever seen and is worthy of the highest of praise. I commend Lieutenant O'Kane for being an inspiration to the ship.

(Endorsement to third war patrol report)

Serial 0198   Care of Fleet Post Office,
                              San Francisco, California,
                              February 12, 1943

CONFIDENTIAL
COMSUBPAC PATROL REPORT NO. 138
U.S.S. *Wahoo* – THIRD WAR PATROL

From: The Commander Submarine Force, Pacific Fleet.
To: Submarine Force, Pacific Fleet.
Subject: U.S.S. *Wahoo* (SS-238) – Report of Third War Patrol.

Enclosure: (A) Copy of Subject War Patrol.
           (B) ComSubRon 10 conf. ltr. Serial 011 of February 8, 1943

     1. The Commander Submarine Force, Pacific Fleet, takes great pleasure in commending the Commanding Officer, Officers, and crew of the *Wahoo* on an outstanding war patrol. This patrol speaks for itself, and the judgment and decisions displayed by the Commanding Officer were sound.

     2. All attacks were carried out in a most aggressive manner, and it clearly demonstrates what can be done by a submarine that retains the initiative.

     3. The *Wahoo* is credited with inflicting the following damage on the enemy:

SUNK

1 destroyer	*Asashio* Class	– 1,500 tons
1 freighter	*Dakar Maru* Class	– 7,160 tons
1 freighter	*Arizona Maru* Class	– 9,500 tons
1 tanker	*Manzyu Maru* Class	– 6,520 tons
1 transport	*Seiwa Maru* Class	– 7,210 tons
	Total:	31,890 tons

U.S.S. *Wahoo* (SS-238)

J. H. Brown, Jr.,
Acting.

DISTRIBUTION
(1M-43)
List III: SS
Special
P1(5), EN3(5), Z1(5),
Comsublant (2), X3(1)
Comsubsowespac (2)
Subschool NL (2)
Comtaskfor 42 (2)

** signature **
E. R. SWINBURNE,
Flag Secretary

Patrol Four, 23 Feb 1943 – 6 Apr 1943

Subject: U.S.S. *Wahoo* – REPORT OF FOURTH WAR PATROL.
(Period from February 23 to April 6, 1943)

PROLOGUE:

    Arrived Pearl on February 7, 1943 from Third War Patrol.
    Commenced refit by tender, relief crew and ship's force. Shifted 4" gun from aft to forward and mounted a third 20-mm gun on the former 4" gun foundation. Completed refit on February 15, 1943.
    Readiness for sea February 17, 1943. Conducted training February 17 to 19 inclusive. Dry-docked at SuBase Pearl February 21, 1943 for emergency repairs to No. 5 torpedo tube shutter. Cleaned and painted bottom. Undocked ship February 22, 1943.

1. NARRATIVE:

February 23:
    13:00(V-W) Underway from Pearl for patrol area via Midway. With surface escort until dark.

February 23-27:
    En route Midway encountering generally rough weather with mostly head seas. Conducted daily dives and training. Sighted several friendly planes en route.

February 27:
    06:00(Y) Picked up air escort on 30 mile circle bearing east from Midway.

U.S.S. *Wahoo* (SS-238)

08:30(Y) Moored starboard side to the port side of U.S.S. *Tarpon*, at SuBase Midway.

14:30(Y) Departed Midway for patrol areas having taken on 16,000 gallons fuel oil and 2,500 gallons fresh water.

Crossed International Date Line.

February 27 to March 11:

En route to patrol areas conducting daily training dives, fire control drills and battle surface drills. Had the unique experience of making passage from Pearl to inside of the China Sea without sighting a plane and consequently made the entire trip on the surface. The seas were generally rough and from ahead. Had to slow to one engine speed several times, because of excess fuel consumption per mile.

During the first torpedo control drill after leaving Midway, the gyro-setting indicator regulators were found to fail intermittently. For ten days Lieutenant R. H. Henderson, spent practically every moment when off watch in tracing out these troubles, finally locating them in loose connections and in improperly adjusted overload relay micro-switch. Through his untiring efforts the equipment was placed in proper operating condition prior to entering the area. He is deserving of the highest praise. It is gratifying to have a torpedo officer of his calibre aboard.

March 11:

01:10(I) Entered assigned area.

06:10(I) Commenced submerged patrol in assigned area and along the Nagasaki – Formosa shipping route. Seas were flat calm.

March 12:

During the night sighted many lighted sampans, which were always in pairs.

05:55(I) Dived in the Shimonoseki – Formosa trade routes, hoping the "beauty" the U.S.S. *Sunfish* hit would limp through today. It was perfect approach weather.

Our plan of operation is to spend a day in each of their known shipping routes while we work our way up north where we hope to locate the route where the heavy traffic from the Yellow Sea flows into the Inland Sea via Shimonoseki.

13:42(I) Sighted masts. Conducted approach only to identify two steam driven sampans about 500 tons each.

17:28(I) Sighted small (60 foot) motor sampan.

Sighted numerous lighted sampans during the night and kept clear.

March 13:

06:00(I) Dived with Maoa To light, just off the Southwest coast of Saishu To, bearing 358° T., distant 6 miles where the Shanghai – Shimonoseki traffic could pass. Also some Yellow Sea traffic could round this corner.

07:00(I) Sighted another small motor sampan.

08:14(I) Sighted smoke. Commenced approach, which lasted almost five hours. The closest we could get was about 8,000 yards. Finally abandoned the approach. A peculiar mirage prevailed. As far as we could tell it was a small Inter-Island type steamer. It was either acting as a smoking decoy and patrol boat or it was trawling. We nicknamed it *"Smoky Maru."*

16:40(I) (FIRST ATTACK). The same *Smoky Maru* headed directly for us. Went to battle stations and made approach. At 17:04(I) fired one torpedo from a stern tube at 1,000 ton ship, range 1,000 yards, 90° port track, speed 12 knots. Missed a few feet ahead of target. After our long chase this morning and being anxious to shoot something, we let him have just one. He was the type of target worth one torpedo if you sink him, but not worth two torpedoes under any conditions.

The torpedo was set to run at five feet. The sea was light (condition 2), however it is believed the torpedo ran shallow. It was seen to porpoise just ahead of the target. It is possible the target did not sight it, because afterwards he held a steady course and speed. Miss was due to error in estimating masthead height.* We guessed 75 feet. Actually it was about 55 feet. This was determined by timing the run of the torpedo when it broached just ahead of the target. Target similar to U.S.S. *Gudgeon*'s sketch of the "Q" ship but without any guns. His turn count gave him 10 turns per knot. Other boats in this area will no doubt sight this type of ship in the future.

18:15(I) Just as the target was going over the horizon another *Smoky Maru* came out to relieve the watch. We avoided.

Sighted many lighted sampans during the night.

March 14:

06:00(I) Dived with Kakyo To light bearing 000° T., distant 3 miles, in position to intercept some of the Yellow Sea traffic which rounds the corner for Shimonoseki, especially traffic from Tsingtao.

06:45(I) Sighted another *Smoky Maru*. He acted as if he was patrolling. He was towing nothing, yet his speed was five knots or less on various

---

* Range was estimated using the stadimeter built into the periscope. This worked by placing the waterline of a ghost image of the target on the true image's masthead, which would indicate the angle between the waterline and masthead. Since the lines of any given angle will touch both ends of a vertical line crossing them only at a particular distance, the stadimeter can take the angle and masthead height and display it as a distance in yards. If the masthead height entered into the stadimeter is incorrect, the range will also be incorrect.

U.S.S. *Wahoo* (SS-238)

courses. The sea was flat calm. The temperature had dropped from 68° to 48° over night.

08:04(I) *Smoky Maru*, after making a wide circle, speeded up to 10 knots. Sighted smoke. Commenced approach. During the approach had as many as five *Smoky Marus* in sight. It certainly looked as if they were acting as decoys trying to sucker us away from a good-sized target that might be smoking.

10:15(I) Abandoned approach after establishing all targets as too small for torpedo fire. Some of these vessels remained in sight during the entire day.

Sighted many lighted sampans during the night.

March 15:

06:00(I) Dived with Hempun To bearing 355° T., distant 11 miles. This was believed to be the route taken by a large volume of the Yellow Sea traffic to Japan. Visibility had slightly decreased with a light haze.

14:14(I) Sighted small patrol or gun boat, range about 8,000 yards. When we swung ship to approach course, we lost him in the haze and were unable to regain contact.

Sighted many lighted sampans during the evening.

March 16:

01:50(I) Radar contact 10,000 yards.

02:00(I) Radar contact 3,200 yards and immediately sighted vessel resembling a destroyer with a very sharp angle on the bow and with moon in back of us. Dived and pointed own ship towards target for a possible "down the throat" defensive shot. Lost sight of the target when the range was about 1,400 yards. The target was not a destroyer, but another *Smoky Maru*.

03:20(I) Surfaced when SJ radar failed to pick up anything.

06:00(I) Dived with Chu To light bearing 138° T., distant 11 miles.

06:20(I) Sighted another *Smoky Maru*. He was making radical and frequent zigs at 7 knots speed.

This Maikotsu Suido is definitely not a good place for submarine attacks. It is shallow, with islands and shoals everywhere. However, we considered it worthwhile to reconnoiter to see where this Yellow Sea traffic is located. We call this channel "Sampan Alley."

Sighted several lighted sampans during the night.

19:40(I) Upon surfacing set course North and, when we crossed the path of the ships sighted by the U.S.S. *Haddock*, we changed course and followed this track heading us for the proximity of Shantung Promontory.

Our SJ radar went out of commission during the night. We have no technician aboard, but Lieutenant C.C. Jackson II and our leading radioman

U.S.S. *Wahoo* (SS-238)

have been relieved of all duties, while concentrating on this valuable instrument.

March 17:
Did not dive this morning. Visibility was excellent and sea calm. Had some difficulty in dodging all of the junks and trawlers to prevent being sighted.

08:00(N) Dived on what we thought was a plane contact. After talking it over, considered contact was very likely a flight of three geese. Stayed submerged while a few junks got out of sight.

10:00(H) Surfaced.

10:55(H) Dived. Too many trawlers and junks were in sight to dodge all of them. This area appeared to be a shipping route, too. A half dozen trawlers remained close to us the remainder of the day.

18:35(H) Surfaced.

Many lighted sampans sighted during the night.

March 18:
04:55(H) Dived with Shantung Promontory light bearing 231°, distant 19 miles. The weather started out hazy and finally ended up with a thick fog. Remained submerged while we worked on SJ radar. Took a few soundings. Weather cleared up at the end of the day. Radar is back in commission. Congratulations to Lieutenant C.C. Jackson II and J.P. Buckley, RM1C.

Upon surfacing set course for Round Island light off the entrance to Dairen. We are bound and determined to find some traffic.

March 19:
04:22(H) (SECOND ATTACK). Sighted freighter. Went to full power and gained position ahead, tracking with radar.

04:55(H) Dived when light enough to see through periscope.

05:15(H) Fired one Torpex[*] torpedo at medium sized freighter identified as *Kanka Maru*, 4,065 tons, range 750 yards, 120° port track, speed 9 knots. Hit. After part of ship disintegrated and the forward part sank in two minutes, and 26 seconds. These Torpex heads carry an awful wallop.

05:20(H) Surfaced to see if anyone survived that blow. Lots of debris and a rowboat were observed but no one left to tell on us.

05:30(H) Sighted another ship.

---

[*] Torpex was first used in American torpedo warheads in 1943. A mixture of TNT, RDX (cyclonite and cyclomethylene trinitramine) and aluminum, Torpex is about 50% more effective than TNT alone.

05:35(H) Dived. Ship turned out to be a junk. So commenced submerged patrol off Dairen.

07:55(H) (THIRD ATTACK). Sighted freighter with large angle on the starboard bow. Commenced high-speed approach. We had to run over seven (7) miles.

09:16(H) Fired two Torpex torpedoes at what appeared to be a new freighter or naval auxiliary in ballast with guns forward and aft; similar to the *Tottori Maru*, 5,973 tons, 125° starboard track, speed 9 knots, range 1,800 yards. First torpedo hit under his foremast with a terrific blast, but his bow remained intact, however, we could see a tremendous hole up his side. Second torpedo hit him amidships, but it was a dud. The co-approach officer saw a small plume and both sound operators heard the thud of the dud.

09:21(H) Checked the set up and fired another torpedo. The target maneuvered and avoided.

09:26 (H) Fired fourth torpedo right up his rump. Again the target maneuvered and avoided. Target fired at periscope. We certainly hated to see this one go over the hill. The water is so shallow around here, we cannot afford to tangle with a concentration of patrols. That dud cost us one fine ship plus two other precious torpedoes and a chance to shoot at more targets at this spot.

09:30(H) Continued submerged patrol heading away from the scene of the morning engagements.

Except for one day of fog, the weather has been perfect. Tonight we are patrolling along the route our two victims came in on yesterday. It leads to a light off the Korean coast and just South of Chinnampo. We shall patrol off of this light tomorrow.

No small fishing boats sighted to-night—the first time "no-see!!"

March 20:

03:10(H) Sighted ship. Commenced approach.

04:40(H) Dived.

05:15(H) Broke off the approach when target turned out to be a small patrol or trawler. The visibility was so good that this small craft was sighted at an unusually long range.

Sighted several smoking ships well inshore and over the horizon. We are going over there tonight and patrol off Chosan Man Point tomorrow. The traffic to Chinnampo, a large port, must pass that point, so we hope to have some luck.

March 21:

The currents encountered around this port were really strange, but conformed with those shown on the chart.

05:10(H) (FOURTH ATTACK). O.O.D. picked up ship with a range about 7,000 yards and angle on the bow 30° starboard. Commenced approach immediately. At second observation ship had changed course 60° to his right, putting us on his port bow, so we swung ship again and closed at high speed.

07:00(H) Fired three torpedoes at large freighter identified as *Seiwa Maru*, 7,210 tons, range 1,600 yards, 117° port track, speed 11 knots. Third torpedo hit him amidships and he went down by the bow, attaining a vertical angle, and was out of sight in four minutes. We counted 33 survivors in the water (temperature of water and air 40° F). There was debris for the survivors to cling to. Considered they could last but a couple of hours. Took several pictures.

This was a Torpex head and they really blow a ship to pieces and the sound is terrific to us. Twice a washbasin has been knocked off the bulkhead in the forward torpedo room.

09:30(H) (FIFTH ATTACK). O.O.D. sighted ship, range 13,000 and angle on the bow 5° starboard. Maneuvered for a stern shot.

09:58(H) Fired a spread of three torpedoes at large freighter identified as *Nitu Maru*, 6,543 tons, 87° starboard track, speed 10 knots, range 800 yards. Two Torpex torpedoes hit, one under his bridge and the other under the mainmast.

This ship went down vertically by the bow and was out of sight in three minutes 10 seconds. Had the water been deeper he would have sunk faster, because the bow was resting on the bottom as it sank. Two junks were nearby and they appeared to be heading to pick up survivors. Ordered battle surface to destroy the junks.

10:37(H) Surfaced and found junks fleeing away instead of heading for the survivors. We chased them, but when we were within two miles of the beach and nearing shoal water we broke off the chase. Also we were just outside of a large port and we did not want to invite trouble, so we headed back for the survivors. Decided to hunt for anything worth salvaging and pick up a survivor.

We found four survivors. Two on the bottom of one overturned boat, one on the bottom of another overturned boat and a fourth floating by in a life jacket.

We attempted to pick up at least one of them. They seemed to ignore us entirely. After a few minutes of this indifference we said to hell with them and went after something worth salvaging. Picked up a couple of House Flags which we cannot identify. One large life ring with S.S. *Nitu Maru* – Taruni painted on it and a large book which appears to be a Merchant Marine Manual.

11:38(H) Departed this area at full power and then commenced a surface patrol heading for Shantung Promontory at two-third speed.

March 22:
Patrolling off Shantung Promontory. Weather has freshened up with seas and wind from the northwest and horizon slightly hazy.
07:00(H) Made trim dive and inspected main motor that was noisy. Found loose brush and repaired it.
08:15(H) Surfaced and continued surface patrol now heading for a point off Laotiehshan Promontory which is just around the corner from Port Arthur. We believe we can contact some Chinwangtao traffic here.
14:00(H) Dived upon sighting two power sampans.
14:35(H) Surfaced when they appeared to be trawling, and continued on towards our new patrol area.
There is, no doubt, a lot of shipping in this area, but one must find it to sink it. We believe we are heading for a good spot.
There is never much water in this "wading pond" known as the Yellow Sea. We have to be careful with our angle on dives to keep from plowing into the bottom. Aircraft and patrols have been scarce, because we are in virgin territory, however, she "ain't" virgin now and we are expecting trouble soon. We hope to get at least four more ships and then expend our gun ammunition on our way home.
We have sighted lots of fishing junks, sampans, trawlers, etc., but only a few cargo-carrying junks.

March 23:
00:43(H) Sighted small ship with sharp angle on the bow and dove. Commenced approach, lost our target in haze.
03:05(H) Surfaced and continued toward our patrol station.
Laotiehshan Channel can be also called "Sampan Alley." We were literally surrounded by them. Strongly believe the ship we just dived for was a junk, because after surfacing we saw a junk that looked like our target.
04:10(H) (SIXTH ATTACK). Sighted small freighter and commenced approach tracking by radar. Checked his course and speed and attained position ahead.
04:30(H) Dived continuing approach.
04:43(H) Fired one TNT torpedo at medium sized collier, identified as *Katyosan Maru*, 2,427 tons, range 1,000 yards, 88° port track, speed 8 knots. Hit collier just under the bridge. The ship was immediately enveloped in a screen of coal dust. She settled fast and slowed down.
04:57(H) Surfaced to head for our patrol point. It was now the crack of dawn and we had about ten miles to go. The collier we had hit thirteen minutes earlier was not in sight.
05:35(H) Dived when it was getting so light we believed we might be sighted from the beach. Since this collier appeared to be about the same size

as the one we sank the other day we had decided to hit her with a TNT warhead to see if we could obtain a comparison. Our conclusion is that all TNT warheads should be converted to Torpex, because they cannot compare to Torpex. Torpex has the necessary force to sink ships.

09:23(H) Sighted a *Smoky Maru* near Promontory, about ten miles away. He was probably going to try and gain some face for the "Nips."

10:03(H) *Smoky Maru* must have dropped a depth charge. Something like a far-away depth charge was heard.

10:04(H) *Smoky Maru* dropped second depth charge.

10:31(H) Third depth charge.

10:32(H) Fourth depth charge.

18:30(H) Surfaced and set course for a point a little to the northwest of Round Island, which is off Dairen. We feel the shipping will avoid coming into Dairen direct and will attempt "an end run."

Had to run the gauntlet again as we passed through "Sampan Alley."

They have secured the light on Round Island, so we know they are re-routing their traffic.

March 24:

05:05(H) Sighted single float type airplane.

06:45(H) Sighted smoke and commenced approach. If we took the normal approach course, the target and *Wahoo* would end up behind the breakwater at Dairen. Knew we could not close the target sufficiently for an attack, but we closed at high speed just to check his course and position.

13:30(H) Established the "Nips" end run route and commenced heading for it. This freighter was between 4 and 5,000 tons. She passed 16,000 yards ahead of us. We hope to get a couple of ships over on this route within the next day or so and before the "Nips" learn of our presence.

19:24(H) (SEVENTH ATTACK). Radar made contact at 10,000 yards, bearing about 090° T. This was considered fine radar and operator performance as the contact was noticed on the same bearing and just short of land contacts.

This indicated to us that we had a ship and that we were right on its track and we were about to stop their end run play. Commenced surface approach tracking by radar and maneuvered for a stern shot. It was quite dark.

19:49(H) Fired a spread of three torpedoes at a large tanker with engines aft, identified as *Syoyo Maru*, 7,499 tons, range 1,700 yards, track 80° starboard, speed 12 knots. The first two torpedoes had premature explosions at end of 18 second run.

Third torpedo missed.

19:55(H) Fired fourth torpedo and it missed.

20:00(H) The tanker let go several 4 or 5 inch rounds at a range of about 3,000 yards, using the Nips' famous flashless powder. One of the shells landed directly ahead us and burst with a loud bang. We dived and tracked the target.

Here again faulty torpedoes frustrated an attack, wasted four valuable torpedoes that we have carried over 5,000 miles, almost caused the *Wahoo* to be destroyed, and allowed the target the time to open up on its radio and frustrate our newly discovered, fertile, shipping route.

20:14(H) (EIGHTH ATTACK). Surfaced after fourteen minutes of ducking target's shots. He was still shooting, but it must have been at random as he had not seen us the past fourteen minutes. Went ahead full power to get up ahead of this fellow quickly or we would both end up in Dairen Harbor.

20:54(H) When were well ahead and had the target in the middle of a rising moon we dived.

21:22(H) Fired a spread of three torpedoes at target 1,200 yards range, 90° starboard track, target speed 10 knots. Second torpedo with TNT head, hit him in the engine room. He sank in 4 minutes 25 seconds, going down by the stern. The target was loaded to the gills with fuel oil.

It is interesting to note that when we tracked this target at slow speeds we get one target speed and when we were making high speeds, we obtained another target speed. The slow speed tracking is more accurate. This is caused by pitometer log inaccuracy.

21:34(H) Surfaced and headed south for another likely spot off O To Light.

March 25:

01:57(H) (NINTH ATTACK). Sighted ship. He had a green light burning constantly which appeared in every respect to be his starboard sight [side?] light. Three hours prior to this our SJ radar training gear jammed and we were still trying to repair it when this contact was made. Consequently had to conduct approach without radar.

The moon was bright, so maneuvered for a favorable position ahead.

03:55(H) Dived and commenced submerged approach.

04:36(H) Fired a spread of two torpedoes at a medium sized freighter, later identified as the *Sinsei Maru*, 2,556 tons, range 1,300 yards, 87° starboard track, speed 8-½ knots. First torpedo exploded prematurely at the end of 26 second run. Second torpedo exploded prematurely at the end of a 49 second run and about fifty yards short of target.

04:44(H) (FIRST GUN ATTACK). Battle surfaced. First 4-inch shot hit target in after deckhouse at 3,800 yards range. Closed in on target and raked him with 20-mm and holed him with almost 90 rounds of 4-inch. Target caught fire in several places. Her lifeboat was dangling from the forward davit. Passed about twelve survivors in the water all sort'a chattering. The crew yelled to the survivors, "So Solly, please."

05:10(H) (SECOND GUN ATTACK). Lookout reported ship on the horizon. Proceeded at flank speed to investigate, leaving first freighter on fire and listing. Upon closing found target to be a neat little diesel driven freighter quite similar to *Hadachi Maru*, 1,000 tons, but definitely a cargo ship.

05:35(H) Commenced firing on second freighter with 20-mm and 4-inch. He caught fire several times, but the fire was extinguished by her crew or it went out on its own accord. She speeded up to about 13 knots and appeared to be trying to ram the *Wahoo*. We had no trouble in keeping clear. A member of her crew was in the foretop waving his arms—maybe he was conning ship. A few 20-mm hits in his vicinity caused him to slide down a guy wire like a monkey.

Repeated gunfire soon had her blazing all over and dead in the water. Quartermaster reported first freighter listing badly during this engagement and before cease firing he reported first freighter sinking rapidly, and finally she was seen to sink.

06:14(H) After expending 170 rounds of 4-inch and about 2,000 rounds of 20-mm on these two freighters, proceeded on our course for our patrol point off O To Light.

Anyone who has not witnessed a submarine conduct a battle surface with three 20-mm and four-inch gun in the morning twilight with a calm sea and in crisp clear weather, just "ain't lived." It was truly spectacular.

Our deck took a beating. Practically every blast of the 4-inch would give a hit on the target and a partial hit on the *Wahoo*. The wooden decking would tear and take off with each shot.

06:25(H) Watched freighter sink through No. 1 periscope.

06:40(H) Aircraft contact. Dived. This is bad for us, because it spoils our new hunting ground. The aircraft is bound to have seen the freighter burning and then sink. So remained submerged conducting high periscope observations.

12:22(H) Sighted large passenger freighter with large angle on the bow with range about 16,000 yards. Commenced high-speed approach. Took observations at 8,000 yards generated range. Our set-up checked surprisingly well. Continued high-speed approach. Took another observation when generated range was 5,000 yards. Target had reversed course and the range was about 12,000 yards. It was possible the target sighted us, but we doubt it. We believe he had an air escort or an aircraft hovering our area warned him. Anyway we lost the best target we have seen this trip.

13:45(H) Sighted aircraft. Something was evidently cooking. As our battery was low, we cleared the area on new course at best speed.

14:58(H) Sighted a new destroyer, range about 8,000 yards, angle on the bow 15° port. He searched with his Q.C. Went to 150 feet, and rigged for depth charge. As water was about 30 fathoms deep, did not dare tackle this fellow with only two torpedoes aboard, which from late experience, would likely be prematures. It hurt our pride to have to hide in our shell and crawl away.

16:55(H) Heard one distant explosion. This could have been either a bomb or depth charge. The pinging had ceased after getting very faint. We figured our Dog Dog had no chance of finding us then. He may have picked up one of the two freighters sitting on the bottom and depth charged it.

18:55(H) Surfaced and cleared present area on three engines.

10:12(H) (THIRD GUN ATTACK). Sighted trawler.

10:20(H) Opened up with 20-mm guns and 4-inch on a diesel trawler of about 100 tons. Holed him several times. A few fires started, but was so water-soaked they soon died out. Threw aboard some homemade Molotov cocktails concocted and manufactured by the Midway Marines. They didn't burn well, due probably to the water-soaked wood. This trawler had a nice radio antenna which he probably opened up on.

10:50(H) Departed, leaving the trawler in pretty much a wrecked condition. It was [too] rough to board her. Otherwise we could have had fresh fish and also opened up some sea valves in her.

During this engagement all three 20-mm guns were jammed at the same time. These guns really do jam often. Our cooling tubes prevented several explosions like we had last trip. The guns actually boil all the water out of these tubes. Other boats should get larger tubes.

March 28:

Conducted surface patrol on Shimonoseki – Formosa shipping routes. We have not had a good fix since night of the 25th. Took occasional soundings throughout the day.

12:35(H) Dived on radar contact; did not sight the plane.

13:38(H) Surfaced. Visibility was poor all day.

18:00(H) (FOURTH GUN ATTACK). Sighted two lighted motor sampans (*Fishi Marus*).

18:08(H) Opened up with two 20-mm guns on the two sampans.

18:20(H) Secured 20-mm guns and crews after expending about 500 rounds on each sampan. They did not sink, but they have a lot of holes in them and they are quite wrecked. It was still too rough to go aboard for a mess of fresh fish. Our mouths watered at such a possibility.

March 19:

02:55(H) (TENTH ATTACK). Sighted ship, and commenced radar tracking.

04:00(H) Dived when we had gained a favorable position ahead and it was light enough to see the target through the periscope.

04:16(H) Fired a spread of two torpedoes at fairly large freighter identified as *Kimisima Maru*, 5,193 tons, range 900 yards, 90° port track, speed 8-½ knots. First torpedo hit under his mainmast, which was our point of aim, and completely disintegrated everything abaft of his stack. The forward section sank two minutes and thirty-two seconds later. The torpedo was set at 15 feet due to rough seas. This was a Torpex head and it is believed was an influence explosion.

The target made a lot of noise as she sunk and broke up. We all could hear it through the hull. The second torpedo was aimed at the foremast. It missed, because the first torpedo stopped the foremast in its tracks!!!!!!!

04:26(H) Surfaced and headed for our base. All torpedoes expended.

Made surface transit through Colnnett Strait during daylight. We had a lot of small craft and a small freighter in sight, all at the same time. They did not bother us and we kept right on going.

07:40(I) Departed out of area.

08:45(I) Dived on radar contact. Did not sight the plane.

09:35(I) Surfaced.

19:35(I) Good SJ radar contact at 9,200 yards. Did not investigate nor did we sight anything.

March 30:

18:27(I) Good contact on the SJ radar at 9,800 yards. Did not investigate nor did we sight anything.

19:10(I) An unusual swell washed over the bridge and flooded the main induction. Water entered the maneuvering room through the auxiliary induction causing water to partially flood the main cubicle. Several zero grounds were created, which in turn started many small fires making the control station untenable by a very caustic smoke. All main propulsion stopped and out of commission. All forced ventilation stopped.

Immediately opened all battery cut-out switches, which stopped the fires. Idled one engine while we took a suction through the after torpedo room hatch. This cleared the maneuvering room of smoke in a hurry, but we took quite a bit of water in the torpedo room.

Commenced clearing up grounds in main control cubicle and clearing main induction of water.

The following parts were burned up or damaged and will have to be replaced during the next refit period:

3 Generator rheostat clutch switches.
1 Generator trip switch.
2 Generator rheostat field contactors.
100 feet of wiring.

20:40(I) Went ahead standard speed on port shaft with No. 2 main engine on propulsion.

March 31:
04:42(I) Went ahead standard speed on both shafts. No. 1 engine available for propulsion.
08:33(I) Dived on plane contact. It was flying very low, distance about 4 miles. Radar did not pick it up. Unfortunately we were right on the route between Tokyo and the Bonins. The sky was heavily over-cast, with a low ceiling. This plane was probably piloting down the chain of the Southern Islands.
09:12(I) Surfaced.
11:00(I) All main engines available for propulsion except No. 3.
12:30(K) Another plane contact, distance about 12 miles by radar. Did not sight it. This plane was in the same groove as the other one, this morning. We did not dive.

April 2:
09:10(K) Sighted sail on the horizon. After closing it a bit we could see a single sail and a long hull. Believed this to be a patrol disguising himself or economizing on fuel. We were several hundred miles east of the Bonins and no sail boat had any business in these parts. Position Latitude 31°-30' N; Longitude 150°-25' E.

April 6:
10:30(Y) Arrived U.S. Submarine Base, Midway Island. Had second unique experience of this patrol of surfacing in middle of Yellow Sea, on March 25th and proceeding on surface from that date until arrival at Midway, with only short trim dives, one submerged attack and two ducking for plane contacts.

2. WEATHER.

The weather was generally crisp and clear except when southerly winds caused fog.

3. TIDAL INFORMATION.

The tides and currents conformed with the information shown on the charts and contained in sailing directions.

U.S.S. *Wahoo* (SS-238)

4. NAVIGATIONAL AIDS.

On entering the areas all Navigational lights shown on chart which were encountered were burning with proper characteristics. However, Round Island light was extinguished after the third attack in that area, and it is presumed others were put out also. On March 29 Kusakaki Shima Light was burning, but in view of the sinking twenty-six miles from it that morning it has probably been doused.

5. ENEMY SHIPS SIGHTED:

Date:	Time:	Position:	Course:	Speed:	Type:
3/14/43	17:04 I	Lat 32°-57 ¼' N Long 126°-11' E	104°	12 knots	1-1,000 ton AK
3/15/43	14:15 I	Lat 34°-04 ½' N Long 122°-18 ½' E	270°	High	Small patrol or gun boat
3/19/43	05:15 H	Lat 38°-29 ½' N Long 122°-18 ½' E	305°	9 knots	1-4,065 ton AK Nanka Maru
3/19/43	09:16 H	Lat 38°-27 ½' N Long 122°-33' E	295°	9 knots	1-AK or Naval Aux Tottori Maru 5,973 tons
3/21/43	07:00 H	Lat 38°-10 ½' N Long 124°-32 ½' E	000°	11 knots	1-AK 7,210 tons Seiwa Maru
3/21/43	09:58 H	Lat 38°-04 ½' N Long 124°-32 ½' E	357°	10 knots	1-AK 6,543 tons Nitu Maru
3/23/43	04:43 H	Lat 38°-37 ¼' N Long 121°-01 ¼' E	138°	8 knots	1-AK 2,427 tons Katyosan Maru
3/24/43	12:49 H	Lat 38°-47' N Long 122°-16 ½' E	265°	10 knots	Freighter passed 16,000 yards ahead 4-5,000 ton
3/24/43	19:49 H	Lat 39°-01' N Long 122°-24 ¼' E	263°	12 knots	1-AO 7,499 tons Syoyo Maru
3/25/43	04:36 H	Lat 38°-12 ½' N Long 123°-24' E	343°	8.5 knots	1-AK 2,550 tons Sinsei Maru
3/25/43	05:10 H	Lat 38°-10' N Long 123°-26' E	Various	13 knots	1-AK 1,000 tons similar to Hadachi Maru
3/25/43	12:22 H	Lat 38°-01' N Long 123°-36' E	300°/ 120°	10 knots	Passenger Freighter large
3/25/43	14:58 H	Lat 37°-55 ½' N Long 121°-01 ¼' E	300°	12 knots	New type DD. It was echo-ranging
3/23/43	04:16 H	Lat 30°-25 ½' N Long 129°-41 ½' E	080°	8.5 knots	1-AK 5,193 tons Kimisima Maru

NOTE: Sampans, Junks, Trawlers and other fishing craft were encountered daily. As many as twenty being in sight at one time from the bridge. In general those off the Korean coast and in the Shantung Promontory and in

U.S.S. *Wahoo* (SS-238)

the Gulf of Pohai were darkened and under sail. Areas listed below held particularly heavy concentrations of fishing craft.

DESCRIPTIONS	LOCATION
Tokara Kaikyo	30°-10' N : 103° E
Shimonoseki – Formosa (trade route)	31°-50' N : 127°-25' E
Maikotsu Suido	34°-20' N : 125°-40' E
Laotiehshan Channel (Loathesome Channel)	38°-35' N : 121°-10' E
Southeast of Shantung Promontory	36°-10' N : 123°-20' E

6. DESCRIPTION OF PLANES SIGHTED.

TIME:	TYPE:	LATITUDE:	LONGITUDE:	COURSE:	ALT:
3/24/43 06:45 H	Float	36°-33' N	122°-12' E	SW	2,500
3/31/43 13:45 H	Large land bomber	36°-27' N	121°-02' E	W	3,500
3/31/43 08:33 I	Large land bomber	31°-05' N	140°-15' E	S	1,000

7. SUMMARY OF SUBMARINE ATTACKS.

Attack No.	1	2	3A
Date	3/13/43	3/19/43	3/19/43
Location (Latitude)	32°-57' N	38°-29' N	38°-27' N
(Longitude)	126°-11' E	122°-19' E	111°-18' N
No. of torpedoes fired	1	1	2
Hits	0	1 Torpex	2 Torpex (1 dud)
Sunk (Tonnage)	0	4,065	0
Damaged or probably sunk	0	0	5,973
Type of target	1,000 ton AK	AK *Nanka Maru*	*Tottori Maru*
Range of firing	1,000	750	1,800
Periscope Depth	64'	60'	64'
Surface Night			
Deep Submergence			
Estimated draft of target	12'	22 ½'	8'
Torpedo depths setting	5'	10'	10'
Bow or stern shot	Stern	Bow	Bow
Track angle	90° P	120° P	126° S
Gyro angles	174 ½°	358 ½°	352° 8 ½°
Estimated Target Speed	12	9	9
Firing interval			1 min. 35 sec.
Spread: amount and kind			Divergent 2°
Time for target to sink		2 min. 26 sec.	

Attack No.	3B	3C	4
Date	3/19/43	3/19/43	3/21/43
Location (Latitude)	38°-27' N	38°-27' N	38°-11' N

U.S.S. *Wahoo* (SS-238)

(Longitude)	122°-18' E	122°-18' E	124°-33' E
No. of torpedoes fired	1	1	3
Hits	0	0	1 Torpex
Sunk (Tonnage)	0	0	7,210
Damaged or probably sunk	0	0	0
Type of target	Tottori Maru	Tottori Maru	Seiwa maru
Range of firing	1,900	2,250	1,500
Periscope Depth	64'	64'	64'
Surface Night			
Deep Submergence			
Estimated draft of target	8'	8'	27'
Torpedo depth setting	10'	10'	10'
Bow or stern shot	Bow	Bow	Bow
Track angle	140° S	180°	117° P
Gyro angles	018 ½°	358 ¾°	359 ¾°; 356°; 358°
Estimated Target Speed	7	7	11
Firing interval			14 sec., 9 sec.
Spread: amount and kind			Longitudinal
Time for target to sink			4 minutes

Attack No.	5	6	7
Date	3/21/43	3/23/43	3/24/43
Location (Latitude)	38°-05' N	38°-27' N	38°-01' N
(Longitude)	124°-33' E	121°-01' E	122°-25' E
No. of torpedoes fired	3	1	4
Hits	2 Torpex	1 TNT	4*
Sunk (Tonnage)	6,543	2,427	0
Damaged or probably sunk	0	0	0
Type of target	AK Nitu Maru	AK Katyosan Maru	AO Syoyo Maru
Range of firing	800	1,000	1,700
Periscope Depth	65'	57'	
Surface Night			Radar
Deep Submergence			
Estimated draft of target	27 ½'	20'	28'
Torpedo depths setting	10'	10'	10
Bow or stern shot	Stern	Bow	Stern
Track angle	87° S	88° P	80° S
Gyro angles	161°; 167°; 168°	004°	182°; 182.5°; 183.5°; 196.5°
Estimated Target Speed	10	8	12
Firing interval	10s; 16s		12s; 21s; 13s
Spread: amount and kind	Longtitudinal		Divergent 1°
Time for target to sink	3 min 10 sec	13 minutes	

Remarks: *First two torpedoes exploded prematurely at end of 18 seconds runs.

U.S.S. *Wahoo* (SS-238)

Attack No.	8	9	10
Date	3/24/43	3/25/43	3/29/43
Location (Latitude)	39°-00' N	38°-13' N	30°-26' N
(Longitude)	122°-16' E	123°-24' E	129°-41' E
No. of torpedoes fired	3	2*	2
Hits	1 TNT	0+	1 Torpex
Sunk (Tonnage)	7,499	0	5,193
Damaged or probably sunk	0	0	0
Type of target	AO	AK	AK
	Syoyo Maru	*Sinsei Maru*	*Kimisima Maru*
Range of firing	1,200	1,300	900
Periscope Depth	60'	62'	64'
Surface Night			
Deep Submergence			
Estimated draft of target	28'	24'	24'
Torpedo depths setting	10'	6'	15'
Bow or stern shot	Bow	Bow	Bow
Track angle	090° S	37° S	90° P
Gyro angles	354 ½°	012°; 018°	353°; 341°
	359 ½°		
	359 ½°		
Estimated Target Speed	10	8.5	8.5
Firing interval	44s; 30s	15s	18s
Spread: amount and kind	Longitudinal	Longitudinal	Longitudinal
Time for target to sink	4 min 25 sec		2 min 32 sec

Remarks: * Both torpedoes exploded prematurely. Runs 26 and 49 seconds.
+ This target was later sunk by 4" gunfire.

Attack No.	11	12	13
	First gun attack	Second gun attack	Third gun attack
Date	3/25/43	3/25/43	3/27/43
Location (Latitude)	38°-13' N	38°-10' N	33°-39' N
(Longitude)	123°-24' E	123°-26' E	125°-23' E
Rounds of 4" Ammunition	90	80	11
Hits, approximate	60	50	8
Sunk (Tonnage)	2,556	1,000	—
Damaged or probably sunk	—	—	100
Type of target	AK	AK	Diesel Trawler
	Sinsei Maru	*Hadachi Maru*	#825
Range	3,800/300	3,000/200	3,000/200
Estimated Target Speed	Various	13	Various

U.S.S. *Wahoo* (SS-238)

Attack No.	14
	Fourth gun attack
Date	3/28/43
Location (Latitude)	31°-39' N
(Longitude)	127°-41' E
Rounds of 4" Ammunition	20-mm 700
Hits, approximate	400
Sunk (Tonnage)	***
Damaged or probably sunk	2 sampans
Type of target	AK
	2 sampans
Range	1,000/50
Estimated Target Speed	Lying to

FIRST AND SECOND SHIPS: Sprayed with 1,000 rounds 20-mm.
" " " " Caught fire stem to stern.
" " " " Sank.
Trawler: Sprayed with 900 rounds 20-mm and 7 Molotov Cocktails.
" Wrecked with 4" gun hits.
Sampans: Wrecked.

8. ENEMY A/S MEASURES

The enemy used gunfire whenever possible as a nuisance factor to keep a submarine down. The night firing of the *Syoyo Maru* was good when our location was disclosed by our prematures. Their flashless powder gives off no more light than a dimmed green flashlight.

9. MAJOR DEFECTS

Periscopes: The periscopes—particularly #2 periscope fogged badly at times. During several approaches this fogging was bad enough to necessitate ducking the periscope to complete an observation. The fogging became heavy enough during the time required to take a bearing (less than 5 seconds) to make an accurate stadimeter range impossible.

This condition has existed in the past on this ship and from conversation with other officers we find that it exists in other ships. It is greatly increased by a differential in temperature, particularly when the water is warmer than the air. Unless definite action is taken to correct this defect, it will continue to be a major handicap to the conduct of a successful submerged attack.

#1 periscope was constantly [used] as a high lookout while on the surface. The training of this periscope was so stiff that it greatly reduced the efficiency of the watch. The overhaul by the tender during the last refit made no appreciable improvement in this condition.

Torpedoes: As noted in the narrative one torpedo failed to explode although it definitely hit the target amidships, and four other torpedoes exploded prematurely. Although at first glance this would appear to be just over a twenty percent failure, we must consider also the additional expenditure of torpedoes involved, for a target worth sinking remains worth sinking as long as she is afloat.

In the case of the *Tottori Maru*, the second torpedo, a Torpex, would undoubtedly have sunk her had it exploded. As it turned out, a new 5,973 ton ship was only damaged and two additional torpedoes had to be expended under unfavorable conditions in an attempt to sink her and to preserve a new found "hunting ground."

The *Syoyo Maru*, had been tracked by radar from 10,000 yards in. Her course and speed were most accurately known. There is every reason to believe that the initial spread of three torpedoes would have sunk her. Yet the first two torpedoes, exploded prematurely, invited counter-attack and necessitated firing an additional torpedo also requiring the expending of three more torpedoes an hour and a half later to sink this 7,499 ton tanker. Again the Japs new found traffic lane was spoiled for further attack.

Against the *Sinsei Maru*, the two prematures necessitated a battle surface with its inherent dangers, and the expenditure of 90 rounds of 4" ammunition to sink the 2,556 ton freighter. That our position was again disclosed by this ship's radio is indicated by the arrival of a plane within two hours. Thus in fact, torpedo failures caused the additional expenditure of six torpedoes and 90 rounds of 4" ammunition.

A conservative estimate is that the *Tottori Maru* and two additional ships could have been sunk if all torpedoes exploded properly, and that one *Smoky Maru* could have been sunk with the 90 rounds of 4" ammunition.

10. COMMUNICATION

Radio reception was good and complete. No attempt was made to use the under-water loop. Difficulty was experienced in clearing a message from in the vicinity of the Bonins. It was receipted for, probably somewhat garbled, by Midway.

    Last serial received   ComSubPac 56 Yoke.
    Last message sent   032130/April.

11. SOUND CONDITIONS AND DENSITY LAYERS

Sound conditions were poor, undoubtedly due to the shallow water. For the same reason no density layers were noted.

U.S.S. *Wahoo* (SS-238)

12. HEALTH AND HABITABILITY

The general health of the crew during the patrol was very good. Climatic conditions were rigorous, cold weather persisting practically all the time. A most adequate supply of heavy clothing was available and was distributed upon sailing. The average temperature while "on station" was approximately 40° F.

There were about seven or eight complaints of colds while underway, only one requiring bed-rest. One case endured, although mildly, from the time of contraction to the end of the patrol, this by a man who was making his first run in submarines. Skin diseases were at a minimum, only one or two cases of athlete's foot or "spic" itch in evidence. One man complained of boils, and was victimized constantly. There was one case of cellutitus, and the complaining patient was confined to his bunk for several days.

There were no injuries other than a few minor cuts and bruises.

Habitability was excellent.

The fresh meats, although kept frozen at 20°, again acquired a most unpalatable taste early in the patrol. This condition persisted on each patrol in spite of every effort to locate and remedy the cause. After the second patrol of this ship shelves and spaces were installed to permit air circulation, a thorough check for possible fuel oil or freon leaks was made, and a fan installed to insure air circulation. Absorption of odors by charcoal was also attempted, but with no apparent results. During each upkeep period the chill and cold room have been completely emptied, scrubbed, and aired with blowers. The situation has been called to the attention of the tender Medical Department, which could offer no solution.

(Endorsements to fourth war patrol report)

FB5-44/A16-3   COMMANDER SUBMARINE DIVISION FORTY-FOUR
Serial 04-B   In Care of Fleet Post Office,
                            San Francisco, California,
                            April 9, 1943.
CONFIDENTIAL

From: The Commander Submarine Division Forty-Four.
To: The Commander Submarine Force, Pacific Fleet.
Subject: U.S.S. *Wahoo*, Report of fourth War Patrol - Comments.

1. The Fourth Patrol of the *Wahoo* covered a period of forty-two (42) days between departure from Pearl Harbor, and return to Midway. Eighteen (18) days were spent in patrol area. Patrol was terminated by expenditure of all torpedoes.

U.S.S. *Wahoo* (SS-238)

2. During this patrol, as on the third patrol of this ship, the outstanding aggressiveness and the magnificent fighting spirit of the captain, officers, and crew were largely responsible for the splendid results obtained.

3. The Commanding Officer displayed great enterprise and excellent judgment in covering his patrol area. The area was covered in a most thorough manner. It is particularly noteworthy that in order to increase the scope of the search a surface patrol was conducted during daylight hours on seven of the eighteen days in the patrol area. This was done after experience indicated that anti-submarine measures in the area were not sufficiently effective to make surface patrol foolhardy.

4. Comments on attacks.

(a) Attack No. 1. Excellent firing position, but range was in error. In this case a[n] echo range to check the periscope range before firing would have been extremely valuable.

(b) Second attack. Approach data on this attack was felt to be very accurate. Therefore, only one torpedo was fired and it did the trick.

(c) Third attack. On attack 3A firing interval was excessive. However, two hits were obtained and target would probably have sunk had second torpedo exploded. On attack 3B a single torpedo fired when spread might have been used to advantage.

(d) Sixth Attack. This ship was not observed to sink but the Commanding Officer feels certain she went to the bottom. He bases this on the belief that the ship was so small it could not have survived the explosion, plus the fact that ship was not sighted when *Wahoo* surfaced thirteen minutes after firing.

(e) Seventh and Eighth Attacks. While frustrated by faulty performance of torpedoes on seventh attack the Commanding Officer would not be denied this valuable target. After an unsuccessful attack, *Wahoo* surfaced, gained a favorable position, and executed a second attack, which resulted in the destruction of the ship.

Attention is invited to the remarks in the patrol report to the effect that tracking the target with submarine making high speed gave different target speed than when tracking at low submerged speed.

(f) Ninth Attack. Here again faulty performance of torpedoes ruined an attack. It will be noted that torpedoes were set on six feet depth on this attack. The sea was calm.

(g) Third Gun Attack. The need for some effective means of setting fire to wooden trawlers and sampans was demonstrated during this attack. During current refit period *Wahoo* plans to have personnel practice throwing buckets of oil with a view to using this method of setting wooden boats on fire during future patrols.

5. Comments on Material.

(a) Main Control Cubicle. As result of the flooding of the main induction small fires were started which resulted in the damaging beyond repair of certain parts of the main control cubicle. Replacement parts must be obtained from Pearl or other outside source before repairs can be completed.

After the water started coming through the auxiliary induction valve it was impossible to close the valve against the flow of water. This valve is located overhead in the starboard side of the maneuvering room aft, very close to after starboard corner of the control cubicle. It is impossible to keep the water from entering the control cubicle if the leakage is at all serious. In order to prevent a recurrence of this casualty the *Wahoo* plans to operate with this valve closed. This will increase the temperature in the maneuvering room but appears to be the only safe procedure until such time as the design of the auxiliary induction valve is changed to correct the present undesirable features.

(b) Periscopes. The difficulty with training number one periscope will be investigated during the refit period and effort will be made to improve the condition.

(c) Torpedoes. Four premature explosions were experienced. It is possible that the second torpedo fired on attack number seven exploded prematurely as a result of passing into the disturbance caused by the first torpedo. It will be noted that both of these torpedoes exploded about the same distance away from the submarine. While the time of the second explosion on attack number nine corresponds quite closely with the expected torpedo run to the target, the Commanding Officer is positive that the torpedo exploded before it reached the target. The target was obscured by extensive spray from the explosion and later observations of the ship before it was sunk by gunfire proved conclusively that it was not damaged by torpedo.

It was the practice on the WAHOO to keep all torpedoes set on depth of ten feet. Deep depths were not set on torpedoes. The seas were calm during firing where prematures were experienced.

The Commanding Officer is convinced that Torpex heads are far superior to TNT. The outward effects of the explosions are much more pronounced and ships sink faster, indicating greater destructive power.

6. The tainting of foods kept in chill and cold rooms may eventually have an adverse effect on the health of the personnel on patrol. Investigations should be conducted to determine how this condition can be improved. It is understood that the *Herring* experienced this trouble and found it necessary to renew the cork lining of the refrigerator because the cement used on the cork affected the taste of the foods.

7. *Wahoo* returned from patrol in very good material condition. While tired and visibly worn by the strain of the patrol, officers and crew were in good health and excellent spirits.

FC5-10/A16-3(FB5-102)   SUBMARINE SQUADRON TEN
Serial 053   In Care Of Fleet Post Office,
                                           San Francisco, California,
                                           April 12, 1943.
CONFIDENTIAL

From: The Commander Submarine Squadron Ten.
To: The Commander Submarine Force, Pacific Fleet.
Subject: U.S.S. *Wahoo* (SS-238), Fourth War Patrol – Comments on.

1. The fourth war patrol was again outstanding and marked by maximum aggressiveness and cool daring. The intelligent planning and sound judgment of the Commanding Officer in making his decisions enabled the *Wahoo* to outsmart the enemy, retain the initiative, and inflict a considerable amount of damage.

2. The patrol extended over a period of 42 days, of which nineteen days were spent in the area. The patrol was terminated on expenditure of all torpedoes.

3. One of the outstanding features of the patrol was the successful penetration into the area by surface cruising alone. The *Wahoo* was almost as successful on return from the area, being forced down on only a few occasions.

4. A total of 24 torpedoes were fired during ten separate attacks. Eight hits were scored, and a ninth hit was a dud. In the attack on the *Kimisima Maru* on March 29, the second torpedo would have been a hit had the first torpedo not done the job too completely. Counting this as a hit, a score of 37.5% was made. Had it not been for the dud and four prematures, this percentage would have been still higher and the tonnage sunk, although considerable, would have been still greater.

5. The gun attacks on the AKs, the trawler, and the sampans were well executed. The percentage of hits obtained and the sinking of over 3,500 tons of cargo-carrying ships by the 4-inch gun crew is most gratifying. The recommendation made by the Commanding Officer for larger tubes for the 20-mm guns is concurred in.

6. The Commander Submarine Squadron Ten takes pleasure in extending a "Well Done" to the Commanding Officer and personnel of the *Wahoo* for a highly successful patrol, during which the following damage was inflicted on the enemy:

U.S.S. *Wahoo* (SS-238)

SUNK

AK (*Nanka Maru*)	4,065 tons
AK (*Seiwa Maru*)	7,210 tons
AK (*Nitu Maru*)	6,543 tons
AK (*Katyosan Maru*)	2,427 tons
AO (*Syoyo Maru*)	7,499 tons
AK (*Kimisima Maru*)	5,193 tons
AK (*Sinsei Maru*)	2,556 tons**
AK (*Hadachi Maru*)	1,000 tons**
Trawler	100 tons**
2 Sampans (approx 50 tons each)	100 tons**

TOTAL SUNK: 36,693 tons
** Sunk by gunfire.

DAMAGED

AK (*Tottori Maru*)	5,973 tons

FF12-10/A16-3(5)/(16)   SUBMARINE FORCE, PACIFIC FLEET
Serial 0484   In Care of Fleet Post Office,
                         San Francisco, California,
                         April 13, 1943.

CONFIDENTIAL

COMSUBPAC PATROL REPORT NO. 164
U.S.S. *WAHOO* – FOURTH WAR PATROL

From: The Commander Submarine Force, Pacific Fleet.
To: Submarine Force, Pacific Fleet.
Subject: U.S.S. *Wahoo* (SS-238) – Report of Fourth War Patrol.

Enclosure: (A) Copy of Subject War Patrol Report.
           (B) Copy of Comsubdiv 44 conf. ltr. FB5-44/A16-3 Serial 04-B of April 9, 1943.
           (C) Copy of Comsubron 10 conf. ltr. FC5-10/A16-3 (FB5-102) Serial 053 of April 12, 1943.

  1. Outstanding in aggressiveness and submarine warfare efficiency, this was the fourth war patrol of the *Wahoo* and the second under its present Commanding Officer. Sinking eight ships, one trawler and two sampans and damaging one other ship, the *Wahoo* continued the outstanding record established on its third war patrol.

U.S.S. *Wahoo* (SS-238)

2. Faulty torpedo performance in the form of prematures subjected the *Wahoo* to dangerous shellfire on a night attack. Another torpedo, a dud, allowed a damaged ship to get away.

3. It is gratifying to note that all of the *Wahoo*'s gun battles were executed only after a careful estimate of the situation was made; each was carried out with military aggressiveness, professional competence and yet free of foolhardy recklessness. These attacks were carried out when they could be made with the submarine having the definite advantage. It is well to remember that our submarines are very valuable and, at the same time, vulnerable targets when gunfire is used as the attacking weapon.

4. Throughput the patrol, the Commanding Officer exhibited excellent judgment in his strategic study of the shipping lanes, thus covering the area efficiently and most productively.

5. The Commander Submarine Force, Pacific Fleet, again takes great pleasure in commending the Commanding Officer, officers and crew of the *Wahoo* on this, their second successive outstanding war patrol. The *Wahoo* is credited with inflicting the following damage to the enemy:

SUNK

1 Freighter (*Nanka Maru* class)	4,065 tons
1 Freighter (*Seiwa Maru* class)	7,210 tons
1 Freighter (*Nitu Maru* class)	6,543 tons
1 Freighter (*Katyosan Maru* class)	2,427 tons
1 Tanker (*Syoyo Maru* class)	7,499 tons
1 Freighter (*Kimisima Maru* class)	5,193 tons
*1 Freighter (*Sinsei Maru* class)	2,556 tons
*1 Freighter (*Hadachi Maru* class)	1,000 tons
*1 Trawler (#825)	100 tons
*2 Sampans	100 tons

TOTAL: 36,693 tons
*Sunk by gunfire.

DAMAGED

1 Freighter (*Tottori* class)	5,973 tons

DISTRIBUTION:
(1m-43)
LIST III, SS
Special:
    P1(5), EN3(5), Z1(5),
    Comsublant (2), X3(1),
    Comsobsowespac (2),

J.H. Brown, Jr.,
Acting

U.S.S. *Wahoo* (SS-238)

Subschool, NL (2),
Comtaskfor 72 (2),
Comsubron 50 (2),
Comsopac (2),
Cinclant (2),
Comtaskfor 16 (1).
/s/
E. R. SWINBURNE,
Flag Secretary.

FF12-10/A16-3(18) SUBMARINE FORCE, PACIFIC FLEET

Serial 067 Care of Fleet Post Office,
San Francisco, California,
1 Feb. 1946

CONFIDENTIAL

From: The Commander Submarine Force, Pacific Fleet.
To: The Chief of Naval Operations
Via: The Commander in Chief, U. S. Pacific Fleet.
Subject: U.S.S. *Wahoo* (SS-238) – Report of Fourth War Patrol.

Reference (a) ComSubPac second cnf. FF12-10/A16-5(5)/(16) serial 0484 of 13 April 1943.

1. In reference (a) U.S.S. *Wahoo* was credited with sinking a total of 36,693 tons of enemy shipping and damaging one freighter of 5,973 tons. The damaged freighter was estimated to be of the *Tottori* class and was attacked in position 38°-27' N. and 122°-18' E. on 19 March 1943. Reliable intelligence, subsequently received, indicates that the *Komi Maru* (very similar to *Tottori* class) of 4,520 tons sank on 20 March as a result of submarine attack. The only other ship attacked in that location on either 19 or 20 March was the *Nanka Maru*, which was sunk by *Wahoo* at 05:13 19 March 1943 and whose loss has been verified by intelligence reports. Accordingly, *Wahoo* is hereby credited with the sinking of *Komi Maru* – 4,520 tons.

2. Accordingly reference (a) is modified as follows:
(a) Strike out all after the word "enemy" in paragraph 3 and substitute therefor:

SUNK

1 Freighter (*Komi Maru*) 4,520 tons
1 Freighter (*Nanka Maru* class) 4,065 tons

U.S.S. *Wahoo* (SS-238)

1 Freighter (*Seiwa Maru* class)	7,210 tons
1 Freighter (*Nitu Maru* class)	6,543 tons
1 Freigher (*Katyoson Maru* class)	2,427 tons
1 Tanker (*Syoyo Maru* class)	7,499 tons
1 Freighter (*Kimisima Maru* class)	5,193 tons
*1 Freighter (*Sinsei Maru* class)	2,556 tons
*1 Freighter (*Hadachi Maru* class)	1,000 tons
*1 Trawler (#825)	100 tons
*2 Sampans	100 tons

TOTAL: 41,213 tons
*Sunk by gunfire.

Frank T. Watkins,
Chief of Staff.

Patrol Five, 25 Apr 1943 – 21 May 1943

Subject: U.S.S. *Wahoo* – REPORT OF FIFTH WAR PATROL
(Period from April 25 to May 21, 1943)

PROLOGUE

Arrived submarine base, Midway Island April 6, 1943 after Fourth War Patrol. On April 7, 1943 commenced refit by Submarine base, relief crew, and ship's force.
During April 21 to 22 conducted training exercises underway.
Ship ready for sea April 25, 1943.

1. NARRATIVE

April 25:
15:00(Y) Departed Midway under air escort for patrol area via Kuril* Islands.
Crossed the International Date Line.

April 27:
Tested new flashless powder during complete darkness. Still our powder is not flashless, but it gives off a diminished flash. This powder cannot be compared with that employed by the Japs against us, on numerous occasions. It is also understood that the British and German's flashless powder is as effective as the Japanese. Furthermore, this powder produces a great

---

* Also spelled "Kurile." Both spellings seem to occur with about equal frequency in various reference works, so Morton's (or his yeoman's) spelling is retained here.

amount of smoke and is considered a handicap to the gun crew when firing with an unfavorable wind.

April 29:
    15:55(N) Slowed to one engine speed (80-90) due to heavy head seas.

April 30:
    10:00(N) Seas having diminished, speeded up to two main engines.
    16:22(L) Upon surfacing from daily submerged drills sheared the shear pins in bow plane rigging mechanism. Stopped; lying to while replacing these shear pins. At 17:20(L), all repairs completed, continued course and speed.

May 2:
    Encountered hail and snow on the morning watch.
    14:23(L) Sighted snow capped mountain peaks on Onekotan Islands of the Kuril Islands.
    16:39(L) Slowed and commenced surface patrol along the Kuril Islands. Will investigate Natsuwa tomorrow in close and submerged.

May 3:
    04:00(N) Dived six miles east of Natsuwa and proceeded to reconnoiter the island. Found a four to five thousand ton freighter broken on beach of Natsuwa opposite Banjo To, apparently a victim of a storm. Fresh appearing paint on the protruding parts indicated that this was a recent wreck. Observed well developed air field consisting of four large hangers with dispersal stowages in back, a large landing field apparently equipped with flood lights, administration buildings, radio station, barracks, etc. This installation is considered comparable to the air station on Eastern Island, Midway. Took several photographs, and plotted positions of outstanding features on chart included as enclosure "A" then cleared island submerged.
    11:05(K) Due to absence of any plane activity surfaced and continued patrol of the Kuril Chain to the southward. The islands observed this far south are barren and completely covered with snow and ice, the installation on Natsuwa being the only indication of any activity.

May 4:
    04:20(K) Dived and proceeded to reconnoiter Noyoro Man on the northeast tip of Etorofu Island where there are sulphur works. The harbor was jammed with float ice and no activity could be observed. As the currents were apparently causing the ice floes to surround us, changed course to southeast to get clear.

ATTACK No. 1

05:25(K) The O.O.D. sighted, through the morning mist, what appeared to be a small ship or patrol, range about six thousand yards, angle on the bow 30° port. This put him on course parallel to the Island Chain. This observation was confirmed by the Commanding and Executive Officers. Five minutes later the ship changed course presenting a zero angle on the bow. The end on view through the mist prevented identification of the ship as other than a small freighter or patrol until the range was 3,200 yards. At this time he was coming out of the mist and the angle on the bow was sufficient to identify him as a larger target and worthy of torpedo fire. On the next observation identified target as an auxiliary seaplane tender and maneuvered for stern torpedo shots.

May 4:

05:58(L) Fired a divergent spread of three torpedoes using stack, forward goalpost and after goalpost as points of aim, range 1,350 yards, 123° starboard track, speed 11 knots. The first torpedo with Torpex head hit between stack and bridge after sixty-second run. The torpedo fired at his forward goalpost evidently passed ahead and the one fired aft must have been erratic or a dud. It is inconceivable that any normal dispersion could allow this last torpedo to miss a 510-foot target at this range. The target tooted her whistle, commenced firing to port, away from the *Wahoo* and then turned away dropping four depth charges. She was observed to have a slight port list, but was evidently quite under control. As this ship, a *Kamikawa Maru* Class XAV-1[*] is capable of 21 knots and did not increase speed, it is considered probable that the one hit limited her speed to the 11 knots determined.

06:36(K) Continued on easterly course to clear ice pack.

13:55(K) Surfaced and continued patrol of Kuril Chain to southward. Ice floes prevented investigating Hitukappu Wan on the south coast of Etorofu.

20:47(K) Sent contact report to ComSubPac concerning the XAV-1.

May 5-6:

Patrolling Kuril Chain.

May 7-8:

Entered area and closed coast at full speed.

04:20(K) Dived 12 miles from coast off Benten Zaki. Observed two freighters with a destroyer or patrol, and a third lonely freighter pass ahead of us well inshore out of range.

---

[*] AV – Seaplane tender.

ATTACK No. 2 A and B

10:39(K) Sighted two ships on northerly course, hugging the shoreline. Commenced approach. Leading ship identified as similar to *Yuki Maru* (5,704 tons) and the second like the *Yomsei Maru* (2,861 tons). The second ship however, was dark gray, fitted with gun mounts and was apparently escorting.

11:15(K) Fired spread to two torpedoes at leading ship, range 900 yards, 107° starboard track, speed 9 knots, followed immediately by a spread of four torpedoes at the escort. The first torpedo hit the *Yuki Maru* under the stack and broke her back. The second torpedo missed ahead. The patrol turned towards and successfully avoided the four torpedoes fired at her, though how she got between those four torpedo tracks will always remain a mystery. As the *Yuki Maru* had sunk, went deep and avoided the patrol at full speed then silent running. None of his depth charges were too close. Observed him from periscope depth and cleared vicinity. Heard considerable distant depth charging or bombing and observed planes searching remainder of the day.

May 8:

Proceeded down coast, skirted fishing fleet, and dived a mile and a half off Kone Saki.

05:12(K) Sighted small ship and made approach. He was running within fifteen hundred yards of the beach, turning into every cove. Broke off attack when he was observed to be too small for torpedo fire.

ATTACK NO. 3

14:13(K) Sighted three ships coming down the coast, commenced approach. The convoy was zig-zagging, and when the range had closed was identified as two escort vessels, similar to the one encountered yesterday, escorting a naval auxiliary similar to the *Kihryu Maru* (9,310 tons).

15:03(K) Fired spread of three torpedoes, range 2,500, 90° port track, speed 10 knots, depth setting fifteen feet. The first torpedo (Torpex) aimed at MOT, prematured after 50 second run half way to the target. The second torpedo aimed at mainmast, and down practically the same track as the first, was evidently deflected by the premature or failed to explode. The third torpedo fired at the foremast hit the point of aim but failed to explode. Both sound operators reported the thud of the dud at the same time that a column of water about ten feet was observed at the targets side abreast of her foremast as the air-flask exploded.

15:10(K) Received first of the series of depth charges expected under these circumstances.

U.S.S. *Wahoo* (SS-238)

May 9:
Proceeded up coast with the intention of closing Kone Saki prior to diving.

ATTACK No. 4 A and B
02:45(K) When 17,000 yards from Kone Saki by SJ range, the radar operator observed two pips, 15,000 and 15,300 yards on the same bearing with the land. Changed course and tracked target group and in the position they would occupy at dawn, dived to 40 feet. Continued tracking by radar and periscope bearings until range was 7,000 yards, then went to sixty feet. The targets were seen, identified as a large tanker and freighter in column, evidently making the night run between ports without escort.

04:40(K) Fired a spread of three torpedoes at tanker identified as similar to *Huzisan Maru* (9,527 tons), range 1,200 yards, 100° port track, speed 10 knots, and immediately thereafter a spread of three more torpedoes at the freighter identified as similar to the *Hawaii Maru* (9,467 tons), range 1,130 yards, 90° port track, speed 10 knots. All torpedoes were set to run at eighteen feet. Just after the fifth torpedo was fired the first hit the tanker amidships breaking her back. She sank by the bow and caught fire aft. The fourth torpedo (a Torpex) hit the freighter under the bridge breaking its back, and the fifth torpedo (TNT) hit her aft. She sank by the stern. Attempted taking some periscope pictures in the meager light; then when both ships had sunk cleared the area to the east.

Heard distant depth charges or bombs throughout the day, and one echo ranging A/S vessel, which passed close on one occasion. Our bathythermograph, which showed a two-degree temperature inversion at 170 feet, gave us extra confidence in our 300-foot depth.

20:20(K) Distant explosions and echo ranging still heard so on surfacing cleared area to northeast to patrol Tokyo-Paramushiru route.

May 10:
Commenced submerged patrol on above route.
10:10(K) Surfaced due to poor visibility and conducted radar search.

May 10-11:
Patrolling Tokyo-Paramushiru route. Nothing sighted except one trawler or patrol, which we avoided.

May 12:
Closed coast and dived two miles off Kone Saki. Numerous sampans and a glassy sea made periscope observation difficult.
06:46(K) Sighted light bomber searching vicinity. Heard several fairly loud explosions.

07:30(K) Sighted another light bomber headed for periscope. Cleared area to the east. Heard numerous distant bombs or depth charges throughout the morning.

ATTACKS No. 5 A, B, C, D.

17:25(K) Sighted distant smoke in the northeast which drew to the south. Commenced approach at standard speed to close the range prior to sunset. Identified target group as two freighters in column, the leading one similar to *Nyoken Maru* (4,021 tons) and the second a huge freighter similar to the *Anyo Maru* (9,257 tons). They were tracked at 9 knots, zig-zagging on base course south, well beyond possible position for submerged attack.

20:05(K) Surfaced and went after convoy at full speed while charging batteries.

20:30(K) Sighted smoke of freighters in clear night.

20:51(K) Picked up freighters on SJ, range 9,400 yards, and commenced working around their stern so that attack could be made with them silhouetted in the setting quarter moon.

22:45(K) Having determined enemy zig-zag plan, and speed as 8.5 knots, dived in position for a "two ship" shot where they would come by in column. As both freighters were loaded set torpedo depth at 18 feet.

23:38(K) Fired spread of two torpedoes at *Anyo Maru*, range 1,200 yards, 95° port track, speed 8.5 knots, and immediately thereafter a spread of two torpedoes at the leading ship, range 1,480 yards, 126° port track, speed 8.5 knots. The first torpedo fired at the mainmast hit. The second torpedo, fired at his stack amidships, is believed to have been erratic or a dud. The target course and speed had been most accurately determined and it is inconceivable that a normal dispersion could cause it to miss. No hits were obtained on the leading ship. The *Anyo Maru* was now observed still going, so waited until the range had opened to 5,000 yards then surfaced and commenced another "end-around." The moon had nearly set, so gained position for surface attack tracking target by radar with TBT bearings as he came in.

May 13:

01:07(K) Fired last remaining bow torpedo at *Anyo Maru*, range 1,800 yards, 90° port track, speed 7.5 knots, and then turned with full rudder and speed for an almost identical stern tube shot. Nothing was seen of the bow torpedo or its wake and the enemy apparently did not know he had again been fired upon.

01:11(K) Fired last remaining torpedo into the *Anyo Maru*, range 1,800 yards, 110° port track, speed 7.5 knots. Some phosphorescence was observed as this torpedo headed to intercept the target. It hit under the bridge with a dull thud, much louder then the duds we have heard only on sound,

U.S.S. *Wahoo* (SS-238)

but lacking the "whacking" which accompanies a whole-hearted explosion. It is considered that this torpedo had a low order detonation. Some sparks were observed on the target above the impact, but he turned away apparently under control, belching smoke. At this time the *Nyoken Maru* which was on our starboard bow opened fire and forced us to dive for six minutes. When we surfaced and closed the *Anyo Maru*, she was lagging a mile behind the *Nyoken*, smoking furiously, and making six knots. We manned the deck gun, but withdrew, quite helpless to stop the cripple, when the *Nyoken* turned and rejoined the *Anyo*.

    02:25(K) Cleared area to east on three main engines.

    03:36(K) Sent message to ComSubPac concerning expenditure of torpedoes.

    22:00(K) Set course for Pearl.

May 18:

    21:05(Y) Sighted four ships on starboard bow, two of them appearing to be destroyers. Believe it to be the convoy for Midway, which Stingray had met earlier. Tracked them on course 290°, speed 7, and sent contact report.

May 19:

    01:30(Y) Received information concerning possible meeting with convoy.

May 21:

    10:00(VK) Arrived Pearl.

2. WEATHER

Excellent weather was encountered throughout the patrol.

3. NAVIGATIONAL AIDS

All lights were burning, but dimmed.

5. ENEMY SHIPS SIGHTED

Date:	Time:	Position:	Course:	Speed:	Type:
5/4/43	05:25(K)	Lat 45°-20' N Long 149°-00' E	060°	11 knots	XAV-1 *Kamikawa Maru* Class
5/4/43	08:32(K)	Lat 40°-04' N Long 142°-54' E	170°	8 knots	2 Freighters 1 Destroyer
5/7/43	08:32(K)	Lat 39°-59' N	350°	8 knots	1 Freighter

U.S.S. *Wahoo* (SS-238)

5/7/43	10:39(K)	Lat 40°-05' N Long 141°-53' E	350°	9 knots	*Yuki Maru* – (*Yoni Maru* gun mounts) – Escort
5/8/43	05:12(K)	Lat 39°-02' N Long 141°-58' E	Various	8 knots	Small ship (1,000 ton)
5/8/43	14:13(K)	Lat 39°-02' N Long 142°-02' E	210°	10 knots	2 Escort Ves (Convt'd AK) 1 Aux (*Kinryu Maru*)
5/9/43	02:45(K)	Lat 38°-57' N Long 141°-49' E	210°	10 knots	Tanker *Huzisan Maru* Freighter *Hawaii Maru*
5/11/43	15:20(K)	Lat 39°-27' N Long 122°-16 ½' E	Various	Slow	Small Patrol Boat or Trawler
5/12/43	17:25(K)	Lat 39°-40' N Long 122°-24 ¼' E 180°	Zigging 8.5 knots 7.5 knots *Anyo Maru*		2 Freighters: *Myoken* &

6. DESCRIPTION OF PLANES SIGHTED.

Date & Time	Type	Latitude	Longitude	Course	Altitude
5/7/43 14:00(K)	Small	40°-00' N	141°-53' E	Circling	Medium
5/12/43 06:36(K)	Light bomber	39°-22' N	142°-09' E	000°	Medium
5/12/43 07:30(K)	Light bomber	39°-22' N	142°-15' E	Various	Medium
5/14/43 12:18(K)	SD Radar 4.5 miles	38°-43' N	151°-43' E	—	—

7. SUMMARY OF SUBMARINE ATTACKS

	1	2A	2B
Attack	XAV	AX	Escort
Time	19:58	01:15	01:15
Date – G.C.T.	3 May 1943	7 May 1943	7 May 1943
Latitude	45°-20' N	40°-05' N	40°-05' N
Longitude	149°-08' E	141°-53' E	141°-53' E
Number and type of torpedoes fired on each attack	3 – XIV3A (1 TNT – 2 TPX)	2 – XIV3A (2 TPX)	4 – XIV3A (3 TPX – 1 TNT)
Hits	1 TPX	1 TPX	0
Number Sunk (Tonnage)	0	5,704	0
Number Damaged or Probably Sunk (Tonnage)	1 Damaged 15,650	0	0
Type of target	XAV-1 *Kamikawa*	AK *Yuki Maru*	Patrol (2,500 Conv)
Range	1,350	900	900
Type of attack: Periscope(P) Surface (S) Night (N) Radar (R)	(P)	(P)	(P)
Estimated draft of target	28'	26'	22'
Torpedo depth setting	12'	15'	15'
Bow or stern shot	Stern	Bow	Bow
Track Angle	123° S	107° S	089° S
Gyro Angle	201°; 203°; 205°	009°; 016°	353°; 353°; 355°; 353°

U.S.S. *Wahoo* (SS-238)

Estimated target speed	11 knots	9 knots	9 knots
Firing interval	11 sec; 9 sec	10 sec	13 sec; 18 sec; 22 sec
Spread – amount and kind	Divergent (Diff pts of aim)	Divergent	Divergent
Was torpedo performance satisfactory?	Yes	Yes	Yes

	3	4A	4B
Attack	AK	AO	AK
Time	05:03	18:40	18:40
Date – G.C.T.	8 May 1943	8 May 1943	8 May 1943
Latitude	39°-02' N	38°-57' N	38°-57' N
Longitude	142°-02' E	141°-49' E	141°-49' E
Number and type of torpedoes fired on each attack	3 – XIV3A (1 TNT – 2 TPX)	3 – XIV3A (2 TPX – 1 TNT)	3 – XIV3A (2 TPX 1 TNT)
Hits	0	1	2
Number Sunk (Tonnage)	0	9,527	9,467
Number Damaged or Probably Sunk (Tonnage)	0	0	0
Type of target	AK *Kinryu Maru*	AO *Huzisan Maru*	AK *Hawaii Maru*
Range	2,900	1,200	1,200
Type of attack: Periscope(P) Surface (S) Night (N) Radar (R)	(P)	(R) (P)	(R) (P)
Estimated draft of target	28'	28'	28'
Torpedo depth setting	15'	18'	18'
Bow or stern shot	Bow	Bow	Bow
Track Angle	090° P	100° P	090° P
Gyro Angle	001°; 002°; 359°	340°; 335°; 334°	357°; 355°; 352°
Estimated target speed	10 knots	10 knots	10 knots
Firing interval	12 sec; 23 sec	17 sec; 13 sec	13 sec; 14 sec
Spread – amount and kind	Divergent	Divergent	Divergent
Was torpedo performance satisfactory?	No*	Yes	Yes

* 1 Dud, 1 premature, 1 erratic run.

	5A	5B	5C
Attack			
Time	13:38	13:38	15:07
Date – G.C.T.	12 May 1943	12 May 1943	12 May 1943
Latitude	38°-40' N	38°-40' N	38°-52' N
Longitude	142°-53' E	142°-53' E	143°-00' E
Number and type of torpedoes fired on each attack	2 – XIV3A (1 TNT – 1 TPX)	2 – XIV3A (1 TPX 1 TNT)	1 – XIV3A TPX
Hits	1	0	0

U.S.S. *Wahoo* (SS-238)

Number Sunk (Tonnage)	0	0	0
Number Damaged or Probably Sunk (Tonnage)	9,257	0	0
Type of target	AK *Anyo Maru*	AK *Myoken Maru*	AK *Anyo Maru*
Range	1,200	1,200	1,800
Type of attack: Periscope(P) Surface (S) Night (N) Radar (R)	(R) (P) (N)	(R) (P) (N)	(R) (P) (N)
Estimated draft of target	30'	24'	30'
Torpedo depth setting	18'	18'	18'
Bow or stern shot	Stern	Stern	Bow
Track Angle	095° P	126° P	090° S
Gyro Angle	190°; 183°	149°; 150°	008 ½°
Estimated target speed	8.5 knots	8.5 knots	7.5 knots
Firing interval	14 secs	15 secs	—
Spread – amount and kind	Divergent	Divergent	—
Was torpedo performance satisfactory?	Yes	Yes	No*

Remarks: * Apparently a dud.

5D

Attack	
Time	19:11
Date – G.C.T.	12 May 1943
Latitude	38°-52' N
Longitude	143°-00' E
Number and type of torpedoes fired on each attack	1 – XIV3A 1 TNT
Hits	0
Number Sunk (Tonnage)	0
Number Damaged or Probably Sunk (Tonnage)	9,257
Type of target	AK *Anyo Maru*
Range	1,800
Type of attack: Periscope(P) Surface (S) Night (N) Radar (R)	(R) (S) (N)
Estimated draft of target	30'
Torpedo depth setting	18'
Bow or stern shot	Stern
Track Angle	110° S
Gyro Angle	162°
Estimated target speed	7.5 knots
Firing interval	—
Spread – amount and kind	—
Was torpedo performance satisfactory?	Yes

8. ENEMY A/S MEASURES

The A/S vessels encountered definitely belonged to the second team. They invariably dropped single charges after attempting to locate us by stopping to listen. Aircraft were used to search in conjunction with these vessels, but arrived too late to be effective on all but one occasion. Only one echo ranging vessel was encountered. Their doctrine seems to require dropping lots of charges and bombs whether they know the submarine's location or not. The A/S vessels observed were all converted freighters with gun mounts fore and aft and characterized by black and white checkered painting forward and after of the bridge structure.

9. MINE SWEEPING OPERATIONS

No mine sweeping operations were observed.

10. MAJOR DEFECTS

No major defects were experienced.

11. COMMUNICATIONS

Radio reception was good and was complete. No difficulty, other than ineffective enemy jamming and spurious transmissions, was encountered in clearing messages to NPM on 8,470 KCs.

The loop-coupling adapter was tested regularly. In the Kurils, NPM could usually be copied at 60 feet, occasionally as deep as 65 feet. In the area results were not as good, reception being difficult at 60 feet.

Last Serial received - Comsubpac Serial 74
Last Serial sent - *Wahoo* 192230 May.

12. SOUND CONDITIONS AND DENSITY LAYERS

Sound conditions in the Kurils were fair to poor. The one ship sighted was not heard until at a very short range. In the area sound conditions were good to excellent. Ships were picked up at 5,000 to 8,000 yards.

In the Kurils water temperature varied from 28 to 34 degrees depending upon proximity to land, ice floes, etc.; however, no pronounced gradients were encountered. In the area water temperature was usually 36 to 40 degrees.

All types of gradients were encountered. In the extreme case of May 9, ten miles east of Kone Saki, the temperature dropped from 37 degrees to 32 degrees in going from periscope depth to 100 feet. It is felt that this alone prevented the echo ranging A/S vessel from gaining sound contact. The bathythermograph was used constantly during dives and was very valuable for predicting the sound conditions existing.

13. HEALTH AND HABITABILITY

Health and habitability were excellent, except for one threatening case of appendicitis during return voyage.

14. MILES STEAMED

En route to Area............................. 2,964
In Area.........................……….. 803
From Area.................................… 3,061

15. FUEL OIL EXPENDED

En route to Area............................ 8.68 gal. per mile.
In Area.........................……. 8.53 gal. per mile.
From Area...............................… 15.80 gal. per mile.

16. ENDURANCE FACTORS

Torpedoes................................. NONE
Others..................................... Indefinite.

17. PATROL ENDED

Patrol ended by orders of ComSubPac after expenditure of all torpedoes.

U.S.S. *Wahoo* (SS-238)

(Endorsements to fifth war patrol report)

FC5-10/A16-3(FB5-102) SUBMARINE SQUADRON TEN
Serial 086   In Care Of Fleet Post Office,
                               San Francisco, California,
                               May 22, 1943.
CONFIDENTIAL

From: The Commander Submarine Squadron Ten.
To: The Commander Submarine Force, Pacific Fleet.
Subject: U.S.S. *Wahoo*, Fifth War Patrol – Comments on.

1. The fifth war patrol of the U.S.S. *Wahoo* was again outstanding in aggressiveness and efficiency. In ten action packed days the *Wahoo* delivered ten torpedo attacks on eight different targets.

2. Although the results were gratifying, faulty torpedo performance cut positive results probably as much as 50%. Such must have been a source of keen disappointment to the Commanding Officer and personnel of the *Wahoo*. Seven hits were observed of twenty-four torpedoes fired, for a score of 29.2%. One of these seven was believed to have been of low order detonation.

3. During this patrol, aggressiveness, determination and fighting spirit of the Commanding Officer, officers and crew again manifested in the excellent results obtained. The Commander Submarine Squadron Ten takes pleasure in congratulating the Commanding Officer and personnel on inflicting the following damage on the enemy:

SUNK

AK (*Yuki Maru* Class)	5,704 tons
AO (*Huzisan Maru* Class)	9,527 tons
AK (*Hawaii Maru* Class)	9,467 tons
TOTAL 24,698 tons	

DAMAGED

XAV-1 (*Kamikawa Maru* Class)	15,650 tons
AK (*Anyo Maru* Class)	9,257 tons
TOTAL 24,907 tons	

FF12-10/A16-3(5)/(16) SUBMARINE FORCE, PACIFIC FLEET
Serial 0484   In Care of Fleet Post Office,
                               San Francisco, California,
                               May 29, 1943.

CONFIDENTIAL

U.S.S. *Wahoo* (SS-238)

COMSUBPAC PATROL REPORT NO. 164
U.S.S. *Wahoo* – FIFTH WAR PATROL

From: The Commander Submarine Force, Pacific Fleet.
To: Submarine Force, Pacific Fleet.
Subject: U.S.S. WAHOO (SS-238) – Report of Fifth War Patrol.
Enclosure: (A) Copy of Subject War Patrol Report.
       (B) None.
       (C) Copy of Comsubron 10 conf. ltr. FC5-10/A16-3
          (FB5-102) Serial 086 of May 22, 1943.

1. The U.S.S. *Wahoo*'s fifth war patrol was the third for the present commanding officer. Typical of the previous two patrols, this one was carried out in the same aggressive and successful manner. These three patrols establish a record not only in damage inflicted on the enemy for three successive patrols, but also for accomplishing this feat in the shortest time on patrol. The Wahoo has sunk a total of 93,281 tons and damaged 30,880 more in only twenty-five patrol days.

2. Once again the Wahoo utilized all the weapons available in conjunction with sound strategic and tactical judgment. This combined with team work of personnel made this fifth war patrol another outstanding example of how to conduct submarine warfare.

3. The Commander Submarine Force, Pacific Fleet, congratulates the Commanding Officer, Officers, and Crew of the U.S.S. *Wahoo* for this their third successive aggressive and successful war patrol during which the following damage was inflicted on the enemy:

SUNK

1 Freighter (*Yuki Maru* Class)	5,704 tons
1 Tanker (*Huzisan Maru* Class)	9,527 tons
1 Freighter (*Hawaii Maru* Class)	9,467 tons
TOTAL 24,698 tons	

DAMAGED

1 Ex-Seaplane Tender (*Kamikawa Maru* Class)	15,650 tons
1 Freighter (*Anyo Maru* Class)	9,257 tons
TOTAL 24,907 tons	

C. A. LOCKWOOD, Jr.

Patrol Six, 8 Aug 1943 – 29 Aug 1943

U.S.S. WAHOO

August 29, 1943
C/O Fleet Post Office,
San Francisco, Calif.

From: The Commanding Officer.
To: The Commander in Chief, United States Fleet.
Via: (1) The Commander Submarine Division 102.
    (2) The Commander Submarine Squadron Ten.
    (3) The Commander Submarine Force, Pacific Fleet.
Subject: U.S.S. *Wahoo*, Report of War Patrol Number Six.
Enclosure: (A) Subject report.
    (B) Track Chart.

1. Enclosure (A), covering the Sixth war patrol of this vessel conducted in the Japan Sea during the period August 8, 1943 to August 29, 1943, is forwarded herewith.

/s/
D.W. MORTON

U.S.S. *WAHOO* – SIXTH WAR PATROL

(A) PROLOGUE TO

Arrived Pearl Harbor, May 21, 1943 after Fifth War Patrol.

On May 22, 1943 Admiral Chester W. Nimitz, USN, Commander-in-Chief, U.S. Pacific Fleet, came aboard and made presentations of awards.

On May 23, 1943 departed for Navy Yard, Mare Island, Cal. and commenced overhaul.

U.S.S. *Wahoo* (SS-238)

On July 11, 1943, completed overhaul.
July 11 to 20th, inclusive post-repair trials and training period.
On July 20, 1943 Captain John B. Griggs, Jr. USN came aboard and made presentation of awards.
On July 21, 1943, departed for Pearl Harbor. Furnished services for surface and air forces the first day while en route.
On July 27, 1943 arrived at Pearl Harbor.
On July 29, 1943 Executive officer and Chief-of-the-Boat transferred to hospital. Diagnosis: Appendicitis Acute. Lieut. Comdr., Verne L. Skjonsby, USN, reported aboard and assumed duties as Executive Officer.
July 30 to Aug. 1, underway for training purposes. Fired three exercise torpedoes.

(B) NARRATIVE

August 2nd:
Departed for patrol Area via Midway. Conducted daily drills.

August 6th:
08:45(Y) Moored alongside U.S.S. *Sperry* at Midway.
17:00(Y) Departed Midway for assigned patrol area.
Conducted daily drills while en route to area.
Crossed International Date Line, skipped Saturday 7th.

August 11th:
20:32(L) Increased to three-engine speed (60-90). Had originally planned to make passage through Yetorofu Straits during the night of August 13th, but fair weather and a following sea has increased our daily distance run. This additional engine will enable us to pass through Yetorofu Straits, during the night of August 12th.
Consider the fuel well expended as it puts us on station a day early.

August 12th:
19:30(K) Slowed to two engines (80-90).
21:47(K) Radar picked up land. Weather foggy, could not sight land.
23:45(K) In center of Yetorofu Straits still conning by radar, visibility zero.

August 13th:
01:45(K) Entered Sea of Okhotsk having completed passage through Strait without sighting land.
02:00(K) Fog lifted, visibility good.

02:35(K) Radar contact, distance 2,600 yards. Could not sight ship, so considered it small and maneuvered to avoid.
04:10(K) Made trim dive. Decided to run on the surface during the day. This will allow us to make passage through La Pérouse Strait tonight. Visibility throughout the day varied many times from zero to unlimited.
17:10(K) Dived on 100 fathom curve, 60 miles east of center of La Pérouse Strait.
18:15(K) O.O.D picked up Man-of-War through periscope, range about 10,000 yards. Went to battle stations and commenced approach. During the early stages of the approach the target looked like a destroyer. But when the range closed to 2,500 yards it was found to be an *Otoroi* Class Torpedo Boat. At this point we reversed tactics and commenced evading. Sound conditions very poor. Temperature of the water dropped 25° with 40 foot change in depth. Final temperature of water was 7° below freezing.

Do not know whether contact with *Otori* Class Torpedo Boat routine or not. He came within a mile of us, stopped, and searched. They could have had tracking information on our SD radar, which had been used during the day.

22:44(K) Sighted smoke dead ahead. Maneuvered to avoid. While avoiding, ship was plainly in sight, but no longer smoked. She was small enough to be the same or another *Otori* Class Torpedo Boat.

August 14th:

01:33(K) Was challenged by shore station on Soya Misaki, range 7 miles. Do not believe he sighted us, as visibility was hazy in that direction. He could have heard us through microphones (doubtful as currents are strong here) or he had some form of radar. Ignored challenge and did not change course or speed (did not want him to suspect us nor did we want to change the sound level).

Both navigational lights were burning and they remained burning after we had been challenged.

02:05(K) Radar contact dead ahead range 5,500 yards. Maneuvered to avoid. He must have been small, as we never did sight him.
04:22(K) Dived.
15:35(K) Surfaced with Rebun Shima bearing 090° T., distance 25 miles. Believe it wise to make a run on the surface for our assigned area, which is about 150 miles south. Will likely reward us with a target during the night. It is a beautiful day.

We arrived in the Sea of Japan in a little over six days with 70,000 gallons of fuel remaining. This is most satisfactory. Good weather combined with Fairbanks-Morse engines is really wonderful.

U.S.S. *Wahoo* (SS-238)

August 14th: All times ITEK (-9 Zone)
    22:17 Sighted smoke over the horizon to the east.
    Commenced tracking on the surface. Upon closing found three freighters heading south. Two of them medium sized and one small. Decided to attack the trailing ship. It could be sunk without the next ship ahead (distance between last two ships, 6,000 yards), knowing what it was all about, thus we could get both ships.

August 15th:
    00:05 Dived for a submerged approached.

ATTACK No. 1
    00:35 Fired one torpedo at medium size freighter, course 205°, speed 7 knots, torpedo run 950 yards, track 96° starboard, torpedo depth setting 10 feet, gyro angle 21° right.
    Miss and no explosion.
    00:55 Surfaced and commenced tracking for another attack.
    01:43 Sighted another ship on a northerly course. It looked like a larger ship and he was heading for us. Broke off the chase on the other freighters and commenced tracking on surface on new target.
    02:06 Dived for a submerged approach.

ATTACK No. 2
    02:22 Fired one torpedo at medium to large freighter, course 030°, speed 11.5 knots, torpedo run 1,150 yards, track 80° starboard, torpedo depth setting 6 feet, gyro angle 180°. Hit at point of aim, but torpedo was a dud and did not explode.
    02:45 Surfaced and commenced tracking for another attack.
    04:15 Dived for a submerged approach. The moon was out, but just setting and dawn had not quite arrived so went in on sharp track to expedite attack while light enough to see.

ATTACK No. 3
    04:18 Fired two torpedoes at same target as in attack No. 2. Course 025°, speed 11.5 knots, torpedo run 700 yards, track 60d starboard, torpedo depth settings 6 feet, gyro angles 345° and 343°. Both missed.
    04:23 One torpedo exploded at end of run.
    In the meantime swung ship and headed directly for target. At the completion of swing target presented a good, up-the-stern shot.

ATTACK No. 4

04:24 Fired one torpedo at same target course 025°, speed 11,5 knots, torpedo run 1,600 yards, track 176°, torpedo depth setting 6 feet, gyro angle 355 ¾°. Missed.

04:27 Torpedo must have broached and exploded before reaching end of run.

This is bad, as it is daylight now and we cannot clear the scene of action and the target will disclose our presence. Damn the torpedoes.

09:30 Echo ranging heard over sound. Soon sighted *Otori* Class Torpedo Boat. Commenced evading. Heard second ship echo ranging, but did not sight it.

12:20 Lost sight of anti-submarine ships.

19:30 While making preparations to surface, sound picked up echo ranging. Soon sighted another *Otori* Class torpedo Boat. Commenced evading. Heard second ship echo ranging but did not sight it.

20:44 Surfaced in a cloudless night and with a full moon. Headed on course 315°, which was directly down moon and cleared the coast. No good hunting tonight with visibility so good. Decided to move over on the Hokkaido-Korea shipping route and spend the night and tomorrow.

August 16th:

17:49 Commenced closing the coast. The moon is bright, but there are a few clouds.

23:39 Sighted freighter heading south. Commenced tracking on the surface. The moon is bright and the target is making such radical zigs, that it is difficult to gain proper position ahead.

August 17th:

00:50 Another contact.
01:03 Dived to avoid detection.
01:17 Surfaced and continued chase.
01:49 Sighted another ship. This one is in a better position to attack, so shifted targets. Commenced tracking latest target.
01:55 Dived for a submerged approach.

ATTACK No. 5

02:26 Fired one torpedo at medium size freighter, course 000°, speed 9 knots, torpedo depth setting 4 feet, gyro angle 359 ¼°. Miss and no explosion. The reason for the large track was to give the torpedo a longer run. We had a perfect position for 600 yard, 90° port track, but our only hit (dud) was at a torpedo run of 1,150 yards, so decided to wait for a similar range.

02:44 Surfaced and cleared the coast. Our tactics are to make night attacks only and clear the coast and rest during the day.

U.S.S. *Wahoo* (SS-238)

04:00 Dived.
11:41 Surfaced for fresh air.
13:00 Dived.
19:38 Surfaced and commenced closing the coast. We plan to shoot low power shots tonight. Maybe the torpedo will have better depth control at low power.
21:42 Sighted freighter heading north. Commenced tracking on the surface.
22:20 Dived and commenced submerged approach.

ATTACK No. 6
22:24 Fired one torpedo at medium size freighter, course 000°, speed 8 knots, torpedo run 1,100 yards, track 77° starboard, torpedo depth setting 4 feet, gyro angle 001 ½° right. Miss and no explosion. This was a TDC controlled low power shot.
22:40 Surfaced. Decided not to chase this ship heading north but wait for a loaded one heading south. Will fire this next torpedo using banjo and zero gyro angle.
23:07 Sighted freighter heading north. He looks a bit larger than the others and partially loaded. Commenced tracking on the surface.

August 18th:
00:05 Dived and commenced submerged approach.

ATTACK No. 7
00:23 Fired one torpedo at medium size freighter course 015°, speed 8.5 knots, torpedo run 850 yards, track 90° port, depth setting 4 feet, gyro angle 000°. Miss and no explosion. Just as we fired a southbound freighter and our target passed each other close aboard; still no hit!!!
01:08 Surfaced and commenced chase after southbound freighter. He is hugging the coast and he is very difficult to see with dark coast as a background. While chasing this ship sighted another one well ahead and away from the coast, so we shifted targets. While tracking on surface, passed two small northbound ships. One looked like a tug and the other a tanker.
03:00 Dived for submerged approach.

ATTACK No. 8
03:11 Fired one torpedo at medium size freighter loaded and on course 165° T., speed 7 knots, torpedo run 1,100 yards, track 45° port, torpedo depth setting 6 feet, gyro angle 225°. Miss and no explosion.

ATTACK No. 9

03:14 Fired one torpedo at same target course 165° T., speed 7 knots, torpedo run 1,100 yards, track 85° port, torpedo depth setting 4 feet, gyro angle 186 ½°. Miss. Torpedo broached at end of 23 second run.
03:17 Explosion. Torpedo must have broached and exploded.
03:30 Surfaced and cleared the coast.
04:07 Dived.
15:15 Surfaced and headed further away from the coast.
Reported to ComSubPac poor performance of our torpedoes.

August 19th:

Received orders from ComSubPac to return to base.
06:47 Sighted ship and commenced surface tracking.
07:58 Dived for submerged approach.
08:48 When about ready to fire at target, her flag was made out to be Russian. Withheld fire and kept out of sight.
09:28 Surfaced and continued toward La Pérouse Strait.
17:07 Dived about 25 miles off La Pérouse Strait. Soya Misaki could be seen through the haze.
19:58 Surfaced and commenced run through the gauntlet.

Again we were challenged, but we ignored them. They definitely could not see us tonight.

Instead of heading directly for Yetorofu Straits, we headed southeast for about four hours. This kept us out of the path of patrols.

August 20th:

08:57 Sighted smoke on the horizon. Upon closing it, found it to be sampan.
09:15 Fired warning shot across bow of sampan. The Japs invariably dive down into their holds when we fire a warning shot. When sampan failed to stop opened up on it with 4" gun and 20-mm guns. After a half a dozen hits with the 4" gun the sampan was a wreck with no sign of life about. Closed sampan to board it. When bow of *Wahoo* was almost touching sampan six members of sampan crew emerged through the wreckage and held up their hands. Six Jap fishermen taken aboard and made prisoners-of-war. Established an armed guard over them. Gave prisoners clean, dry clothes, baths and a round of brandy. Pharmacist Mate examined all prisoners and found only one with a slight shrapnel wound on his knee. None of them can speak English. However, through sign language we were able to learn that five members of their crew had been lost during the engagement. They said that they had come from a port just north of Tokyo and had taken passage through Tsugaru Straits and La Pérouse Straits and they were en route to Onekotan or thereabouts.

U.S.S. *Wahoo* (SS-238)

Prisoners seemed to be grateful for being picked up.
16:39 Sighted smoke on horizon. Commenced tracking on the surface.
16:49 Dived in order to close and take a look at short range.
17:36 Battle surface on another sampan. Fired warning shot across his bow. Again they all dived for the holds.
Opened up with 4" gun and 20-mm guns. Soon had the sampan in roaring flames. Various members of the crew would jump overboard, hide behind their boat and then climb aboard again. None of them ever showed any desire of being rescued.
17:59 While finishing up with present sampan, lookout reported smoke on the horizon.
Commenced surface tracking.
18:14 Dived for closing and to get a look at short range.
19:01 Battle surfaced and fired warning shot across sampan's bow. Again they dived below.
Opened up on sampan with 4" gun and 20-mm guns. After four shots and four hits sampan commenced sinking rapidly.
Went alongside to pick up any willing survivors. A small rowboat was floating. One Jap climbed in the rowboat and several others in the water gave no signs of wanting to be picked up.
These latter two sampans were headed for Yetorofu Island.
19:48 Set course for Yetorofu Straits.
22:00 Completed passage of Yetorofu Straits without sighting land using radar entirely.

August 25th:
Arrived in Midway at 11:07.
Unloaded ten (10) torpedoes.
17:25(Y) Underway for Pearl.

August 29th:
10:35 Arrived in Pearl Harbor.

(C) WEATHER

Good weather was encountered during the entire trip.

(D) TIDAL INFORMATION

Currents encountered were as given in sailing directions. Except that the northerly current in the Japan Sea was stronger than expected, (about 1.8 knots).

U.S.S. *Wahoo* (SS-238)

(E) NAVIGATIONAL AIDS

Soya Mikasi, Noshappu Misaki, and Kanoi Misaki and the Kanoi Misaki lights were observed, showing approximately characteristics given in light lists, but reduced in intensity.

(F) ENEMY SHIPS SIGHTED

No.	Time Date	Lat. Long.	Type(s)	Initial Range	Est. Course Speed	How Contacted	Remarks
1.	16:35(Z) Aug 12	45°-50' N 143°-50' E	— —	— 2,800 yds.	— —	Radar Surface	Evaded
2.	08:15(Z) Aug 13	45°-48' N 143°-42' E	*Otori* Class DD	13,500 yds.	Various	Periscope Submerged	Echo-Ranging Evaded
3.	12.44(Z) Aug 13	45°-43' N 142°-58' E	—	6,000 yds.	—	Binocular Surface	Evaded
4.	16:50(Z) Aug 13	45°-39' N 141°-40' E	—	5,500 yds.	—	Radar Surface	Evaded
5.	12:00(Z) Aug 14	43°-12' N 140°-00' E	4,000T AK 3,200T AK 2,000T AK	20,000 yds. 20,000 yds. 20,000 yds.	205° 7 knots	Binocular Surface	Attack No. 1
6.	16:00(Z) Aug 14	43°-07' N 139°-55' E	6,600T AK	16,500 yds.	025° 11.5 knots	Binocular Surface	Attacks 2, 3, 4
7.	23:45(Z) Aug 14- Aug 15 02:00(Z)	43°-12' N 139°-43' E	2 *Otori* Class DDs	12,000 yds.	—	Sound Submerged	Echo Ranging Evaded
8.	10:00(Z) Aug 15	43°-10' N 139°-35' E	1 *Otori* Class DD 1 Unknown	10,000 yds. ???	—	Sound Submerged	Echo Ranging Evaded
9.	15:00(Z) -18:00(Z) Aug 16	42°-45' N 139°-50' E	4,000T AK 3,500T AK 1,500T AK	10,000 yds. 8,000 yds. 8,000 yds.	180° 7 k 180° 7 k 000° 9 k	Binocular Surface	Attack No. 5
10.	12:30(Z) Aug 16	42°-16' N 139°-43' E	3,500T AK	10,000 yds.	000° 8 knots	Binocular Surface	Attack No. 6
11.	14:30(Z) Aug 17	42°-17' N 139°-43' E	3,000T AK	1,100 yds.	015° 8.5 knots	Binocular Surface	Attack No. 7
12.	16:50(Z) Aug 17	41°-58' N 139°-53' E	4,000T AK	10,000 yds.	165° 7 knots	Binocular Surface	Attacks #8 #9
13.	17:15(Z) Aug 17	42°-05' N 139°-53' E	1,200T AK 800T AT	10,000 yds.	335° 6 knots	Binocular Surface	Sighted Prior #8 #9
14.	21:47(Z) Aug 18	44°-43' N 138°-55' E	3,200T KA	14,000 yds.	255° 6 knots	Periscope Surface	Russian
	22:57(Z)	45°-35' N	36 Ton	9,000 yds.	235°	Periscope	Sunk by

U.S.S. *Wahoo* (SS-238)

15. Aug 19 06:39(Z)	146°-50' E 45°-50' N	Sampan 30 Ton	13,000 yds.	6 knots 240°	Surface Periscope	Gunfire Sunk by
16. Aug 20 07:52(Z)	148°-22' E 45°-47' N	Sampan 35 Ton	12,000 yds.	6 knots 210°	Surface Binocular	Gunfire Sunk by
17. Aug 20	148°-42' E	Sampan		7 knots	Surface	Gunfire

(G) AIRCRAFT CONTACTS

None.

(H) ATTACK DATA

TORPEDO ATTACK FORM				
U.S.S. Wahoo	Torpedo Attack No. 1		Patrol No. 6	
Time: GCT 15:35	Date: Aug 14, 1943	Lat 43°-12' N		Long 140°-00' E
TARGET DATA – DAMAGE INFLICTED				
Description:	3,000 ton AK – last ship in rough column of three. No escort. Visual contact. Full moon over-cast surface visibility good. Estimated range at time of contact 10 miles			
Ship(s) Sunk:	None			
Ship(s) Damaged or Probably Sunk:	None			
Damage Determined by:	—			
Target Draft: 15	Course: 205°	Speed: 7		Range: 950 (at firing)
OWN SHIP DATA				
Speed: 4.5 K	Course: 100°	Depth: 64 Ft		Angle: 009 ½° (at firing)
FIRE CONTROL AND TORPEDO DATA				
Type Attack:	Night, Radar, Periscope. Tracked for period of one hour with radar and TBT bearings then dived for periscope attack. Problem checked precisely on TDC. Fired single torpedo. Miss.			

TORPEDO ATTACK FORM				
U.S.S. Wahoo	Torpedo Attack No. 2		Patrol No. 6	
Time: GCT 17:22	Date: Aug 14, 1943	Lat 43°-07' N		Long 139°-55' E
TARGET DATA – DAMAGE INFLICTED				
Description:	6,000 Ton AK, steaming alone. Visual contact, full moon, overcast sky, surface visibility good.			
Ship(s) Sunk:	None			
Ship(s) Damaged or Probably Sunk:	None			

U.S.S. *Wahoo* (SS-238)

TORPEDO ATTACK FORM			
U.S.S. Wahoo	Torpedo Attack No. 3		Patrol No. 6
Time: GCT 19:18	Date: Aug 14, 1943	Lat 43°-15' N	Long 140°-03' E
TARGET DATA – DAMAGE INFLICTED			
Description:	6,000 Ton AK, same target as attack # 2.		
Ship(s) Sunk:	None		
Ship(s) Damaged or Probably Sunk:	None		
Damage Determined by:	—		
Target Draft: 14	Course: 025°	Speed: 11.5	Range: 800 (at firing)
OWN SHIP DATA			
Speed: 3 K	Course: 269°	Depth: 64 Ft	Angle: 345° 343° (at firing)
FIRE CONTROL AND TORPEDO DATA			
Type Attack:	Night, Radar, Periscope. Tracked target for another hour and a half after previous attack while gaining position ahead. Target data checked with that for initial attack. After diving fired spread of two torpedoes using as points of aim points ¼ length from bow and stern. Both missed.		

TORPEDO ATTACK FORM			
U.S.S. Wahoo	Torpedo Attack No. 4		Patrol No. 6
Time: GCT 19:24	Date: Aug 14, 1943	Lat 43°-15' N	Long 140°-03' E
TARGET DATA – DAMAGE INFLICTED			
Description:	6,000 Ton AK, same target as #2 attack.		
Ship(s) Sunk:	None.		
Ship(s) Damaged or Probably Sunk:	None.		
Damage Determined by:	—		
Target Draft: 14	Course: 025°	Speed: 15.5	Range: 1,200 (at firing)
OWN SHIP DATA			
Speed: 3 k	Course: 025°	Depth: 64 Ft	Angle: 355° (at firing)
FIRE CONTROL AND TORPEDO DATA			
Type Attack:	Night, Radar, Periscope. Swung ship after unsuccessful second attack on this target and as favorable 180° track presented fired single torpedo. Miss.		

U.S.S. *Wahoo* (SS-238)

TORPEDO ATTACK FORM			
U.S.S. Wahoo	Torpedo Attack No. 5		Patrol No. 6
Time: GCT 17:26	Date: Aug 16, 1943	Lat 42°-45' N	Long 139°-50' E
TARGET DATA – DAMAGE INFLICTED			
Description:	4,000 Ton AK steaming alone. Visual contact full moon, scattered clouds. Surface visibility good.		
Ship(s) Sunk:	None.		
Ship(s) Damaged or Probably Sunk:	None.		
Damage Determined by:	—		
Target Draft: 10	Course: 359°	Speed: 9	Range: 1,100 (at firing)
OWN SHIP DATA			
Speed: 3 k	Course: 032°	Depth: 63 Ft	Angle: 007 ¼° (at firing)
FIRE CONTROL AND TORPEDO DATA			
Type Attack:	Night, Radar, Periscope. Tracked target with radar and TBT bearings for 45 minutes prior to diving for submerged periscope approach. Purposely allowed range to open as only previous success on this patrol had been with 1,100 yard run. Fired single shot. Missed.		

TORPEDO ATTACK FORM			
U.S.S. Wahoo	Torpedo Attack No. 6		Patrol No. 6
Time: GCT 13:24	Date: Aug 17, 1943	Lat 42°-16' N	Long 139°-39' E
TARGET DATA – DAMAGE INFLICTED			
Description:	3,500 Ton AK steaming alone. Visual contact, full moon, heavy clouds surface visibility good.		
Ship(s) Sunk:	None		
Ship(s) Damaged or Probably Sunk:	None		
Damage Determined by:	—		
Target Draft: 10	Course: 000°	Speed: 8	Range: 1,200 (at firing)
OWN SHIP DATA			
Speed: 3.5 k	Course: 256.5°	Depth: 61 Ft	Angle: 257.5° (at firing)
FIRE CONTROL AND TORPEDO DATA			
Type Attack:	Night, Radar, Periscope. Tracked target for half-hour on surface, using TBT bearings radar ranges. Dived, continued to track by radar until commencing periscope attack. Fired single shot. Miss. Used low speed in hope of obtaining better depth control.		

U.S.S. *Wahoo* (SS-238)

TORPEDO ATTACK FORM				
U.S.S. Wahoo	Torpedo Attack No. 7			Patrol No. 6
Time: GCT 15:23	Date: Aug 17, 1943		Lat 42°-27' N	Long 139°-43' E
TARGET DATA – DAMAGE INFLICTED				
Description:	3,000 Ton AK steaming alone. Full moon, heavy clouds, surface visibility good. Visual contact.			
Ship(s) Sunk:	None			
Ship(s) Damaged or Probably Sunk:	None			
Damage Determined by:	—			
Target Draft: 10	Course: 015°		Speed: 8.5	Range: 850 (at firing)
OWN SHIP DATA				
Speed: 3	Course: 102.5°		Depth: 63 Ft	Angle: 016° (at firing)
FIRE CONTROL AND TORPEDO DATA				
Type Attack:	Night, Radar, Periscope. Tracked target for an hour using radar ranges and TBT bearings. Set up checked closely on TDC. Set up and fired low speed shot employing angle from MK VIII angle solver. TDC checked exactly with angle solver. Fired single shot. Miss.			

TORPEDO ATTACK FORM				
U.S.S. Wahoo	Torpedo Attack No. 8			Patrol No. 6
Time: GCT 18:11	Date: Aug 17, 1943		Lat 41°-58' N	Long 139°-43' E
TARGET DATA – DAMAGE INFLICTED				
Description:	4,000 Ton AK steaming alone. Moonlight, high clouds, visibility good. Visual contact.			
Ship(s) Sunk:	None.			
Ship(s) Damaged or Probably Sunk:	None.			
Damage Determined by:	—			
Target Draft: 15	Course: 165°		Speed: 7	Range: 1,250 (at firing)
OWN SHIP DATA				
Speed: 3.5 K	Course: 075°		Depth: 55 Ft	Angle: 223.5° (at firing)
FIRE CONTROL AND TORPEDO DATA				
Type Attack:	Night, Radar, Periscope. Tracked with radar and TBT, dived for periscope TDC attack. High speed settings. Missed sixth single shot.			

U.S.S. *Wahoo* (SS-238)

TORPEDO ATTACK FORM			
U.S.S. Wahoo	Torpedo Attack No. 9		Patrol No. 6
Time: GCT 18:14	Date: Aug 17, 1943	Lat 41°-48' N	Long 139°-53' E
TARGET DATA – DAMAGE INFLICTED			
Description:	4,000 Ton AK. Same target as attack #8.		
Ship(s) Sunk:	None.		
Ship(s) Damaged or Probably Sunk:	None.		
Damage Determined by:	—		
Target Draft: 15	Course: 165°	Speed: 7	Range: 1,200 (at firing)
OWN SHIP DATA			
Speed: 3.5 k	Course: 076°	Depth: 55 Ft	Angle: 193.5° (at firing)
FIRE CONTROL AND TORPEDO DATA			
Type Attack:	Night, Radar, Periscope. Target data same as attack No. 8. Torpedo broached after 23 second run.		

Tubes Fired	# 3	# 7	# 2	# 3	# 4	# 1	# 1	# 4	# 8	# 9
Track Angle	96S	80S	60S	60S	176S	150P	77S	90P	45P	85P
Gyro Angle	21R	0	5.5L	4.25L	4.25L	.75L	1.5R	0	45R	6.75R
Depth Set	10	6	6	6	6	4	4	4	6	4
Power	HP	HP	HP	HP	HP	HP	LP	HP	HP	HP
Hit/Miss	Miss	Miss	Miss	Miss	Miss	Miss	Miss	Miss	Miss	Miss
Erratic	No	No	No	No	No	No	No	No	No	No
Mk. Torpedo	XIV	3A	—	—	—	—	—	—	—	—
Serial No.	22791	23255	24390	22769	22798	24506	22764	24505	32612	23026
Mk. Exploder	VI	-1	—	—	—	—	—	—	—	—
Serial No.	6327	6372	6357	6401	17463	6622	6453	7687	6456	5418
Act. Set.	Contact	—	—	—	—	—	—	—	—	—
Act. Act.	None	Dud	None	None	None	None	None	None	None	None
Mk. Warhead	XVI	—	—	—	—	—	—	—	—	—
Serial No.	9674	1475	1438	1081	9572	9641	1600	2619	1461	2329
Explosive	TPX	—	—	—	—	—	—	—	—	—
Fir. Intervl	Single	Single	11 sec		Single	Single	Single	Single	Single	Single
Type Spread	Single	Single	Longitudinal		Single	Single	Single	Single	Single	Single
Sea Cond.	1	1	1	1	1	1	1	1	1	1
Overhaul Activity	------------------------SUBBASE PEARL HARBOR------------------------									
Remarks	Miss	Dud	Miss	Miss	Miss	Miss	Miss	Miss MK VIII Angle Solver	Miss	Miss Broached 23 sec

U.S.S. *Wahoo* (SS-238)

GUN ATTACK REPORT FORM

U.S.S. Wahoo	Gun Attack No. 1		Patrol No. 6
Time: 22:57(Zed)	Date: Aug 19, 1943	Lat 45°-35' N	Long 146°-50' E
TARGET DATA – DAMAGE INFLICTED			
Ship(s) Sunk:	One enemy fishing vessel approximately 36 tons		
Ship(s) Damaged or Probably Sunk:	None		
Damage Determined by:	Observation by ship's company.		

DETAILS OF ACTION (ALL TIMES ZED)

22:57: Sighted smoke on the horizon bearing 110° T, 22° R. Changed course and speed to close. Target turned out to be an enemy fishing vessel.
23:15: Fired warning shot across enemy's bow.
23:26: Commenced firing 4"/50 and 20-mm machine guns.
23:28: Ceased firing, target sinking. At this time six (6) survivors came topside. Decided to take prisoners of war.
A max firing range of 100 yards was used. 4"/50 high velocity ammunition with Mk 30-1 point detonation fuse was used. The "superquick" feature was employed on all nine (9) rounds fired. All hits proved this ammunition quite effective on this type of target. Hits at the waterline caused quick sinking of the enemy. Approximately 120 round of 20-mm ammunition, with a ratio of one (1) HET to two (2) HEI was used in this attack. This machine gun fire started no fires on this target.

GUN ATTACK REPORT FORM

U.S.S. Wahoo	Gun Attack No. 2		Patrol No. 6
Time: 06:39(Zed)	Date: Aug 20, 1943	Lat 45°-50' N	Long 148°-22' E
TARGET DATA – DAMAGE INFLICTED			
Ship(s) Sunk:	One enemy fishing vessel approximately 25 tons		
Ship(s) Damaged or Probably Sunk:	None		
Damage Determined by:	Observation by ship's company.		

DETAILS OF ACTION (ALL TIMES ZED)

06:39: Sighted smoke bearing 033° T., 317° R. Changed course and speed to close target.
06:49: Dived for close observation. Identified as enemy fishing vessel.
07:36: Battle surfaced manning all guns.
07:46: Commenced firing, still pursuing enemy.
07:51: Observed target burning well, ceased firing. Sampan still not in sinking condition so a few more rounds of 4"/50 were placed at her waterline at close range.
07:58: Target sinking fast. No survivors were taken.
Fifteen (15) rounds of 4"/50 high-capacity, point-detonating ammunition with the Mk 30-1 "superquick" feature operating were used. All hits had a devastating effect on the upper-works of the sampan and the hits at the waterline caused rapid sinking.
170 rounds of 20-mm ammunition loaded one (1) HET to two (2) HEI were also used to good effect. They definitely started a good blaze that was stopped when the sampan sunk.
Mean range of firing was 1,000 yards.
*07:45: Fired warning burst across bow.

U.S.S. *Wahoo* (SS-238)

GUN ATTACK REPORT FORM			
U.S.S. Wahoo	Gun Attack No. 3		Patrol No. 6
Time: 07:59(Zed)	Date: Aug 20, 1943	Lat 45°-47' N	Long 148°-42' E
TARGET DATA – DAMAGE INFLICTED			
Ship(s) Sunk:	One enemy fishing vessel approximately 35 tons		
Ship(s) Damaged or Probably Sunk:	None		
Damage Determined by:	Observation by ship's company.		
DETAILS OF ACTION (ALL TIMES ZED)			
07:59: Sighted ship bearing 180° R., shortly after sinking the fishing vessel described in attack #2. 08:14: Dived and tried to close target. 09:01: Battle surfaced manning all guns and pursued target at full power. 09:17: Fired warning shot across bow. 09:19: Commenced firing. 09:30: Ceased firing—sampan sinking rapidly. By the time this ship got to the wreckage the target had sunk. Attempted picking up survivors, no success. Four (4) rounds of 4"/50 high capacity, point detonating ammunition with the Mk 30-1 "superquick" feature operative were used. Four direct hits, several of which were at the waterline, undoubtedly accounted for the rapid sinking of this vessel. Attacks show this ammunition quite satisfactory, at least on this type of target. Fifty (50) rounds of 20-mm ammunition loaded one (1) HET to two (2) HEI was fired, but due to the short time those guns were able to fire at [target] effectively, no conclusive observations could be made concerning their effect in this attack. The mean range of firing was 1,500 yards.			

(I) MINES

None encountered.

(J) ANTI-SUBMARINE MEASURES AND EVASION TACTICS

Several *Otori* Class torpedo Boats conducted a search with their echo-ranging sound gear. 17 KCs was normally used, but one patrol boat used 19 KCs. They dropped no depth charges.

(K) MAJOR DEFECTS AND DAMAGE

The port propeller shaft has a squeal in it at high speed at deep depth. It has a heavy vibration at minimum speed at deep depth. When backing down on the surface it vibrates very badly. At the first available opportunity this shaft should be put in proper working order. Its present condition could be a hazard to the ship during evasive tactics.

(L) RADIO

Radio reception was in general very good in spite of the fact that we were entirely surrounded by land while on station. 450 KCcs and 4115 series were guarded continually and 4235 series at night. Nothing was received on either 450 KCs or 4235 KCs. NPM was copied at periscope depth on the NL loop as far as 2,000 miles from base, under good conditions. Little jamming was encountered. The only transmission made on station was made with no difficulty, NPM answering immediately. Signal strength was five, frequency 8,470 KCs. On the second transmission en route to the base, considerable difficulty was encountered in trying to raise NPM. The message was finally cleared through NPM and NPG.

Last serial received   250843 ComSubPac
Last serial sent   221330 WAHOO

(M) RADAR

S.J. Radar performance was very erratic with maximum ranges on land ranging from 13,800 yards to 28,000 yards. The minimum ranges were obtained under conditions of heavy fog. Overall performance could be considered fair as maximum ranges on targets appeared shorter than on past performance. No lost time.

The S.D. Radar proved useful for navigation and though it was inoperative at one crucial time, the set was very reliable throughout the rest of the trip with land ranges up to 41 miles. Total lost time four hours.

(N) SOUND GEAR AND SOUND CONDITIONS

Sound conditions en route to and on station can be described in one word, terrible. At times the temperature of the water dropped as much as 10 degrees in ten feet at periscope depth. Temperature on one occasion dropped from 58° to 25° from periscope depth to 300 feet. The average temperature differential on station was 23°. Two to three density layers were encountered every time we went deep. The bathythermograph was a great help. During approaches, target screws were heard up to 3,000 yards. The QC sound head was used to "ping" a range, just prior to firing, but sound conditions were so poor that no reliable results were obtained.

(O) HEALTH AND HABITABILITY

Health of the crew was good. A few members of the crew had minor head colds, which were treated with nose sprays, aspirin, and A.P.C. capsules. One case of constipation required bed and rest. One case admitted with diagnosis #2715, tooth unerupted, right inferior third molar. Treated with incision of the gingiva and drained, this was followed with iodoform gauze packing, and sodium perborate mouth-wash every three hours. The third day, sulfathiazole powder was applied with a powder blower, followed by iodoform gauze packing. Total sick days; five. One case suffered from Caries teeth, left superior, second bicuspid, temporary filling was made from Zinc Oxide powder and Eugenol. When applied this relieved all pain. The Commanding Officer suffered slight rheumatic pains two or three times during the patrol, not severe enough to cause him to turn in. The pains were efficiently treated with aspirin grs. 10 and Codeine grs. 1/2. Two doses usually relieved the pain. About six days out from the base, inspection revealed sixteen men infested with Pediculi Pubis*. Due to the small amount of Mercurial Ointment on board, these men were sponged with Diesel fuel oil, followed in one hour with a shower. Another inspection, three days later showed four men still infested. They were instructed to shave and repeat the fuel oil bath. Final inspection, two days later showed the crew to be completely free of the vermin.

Six prisoners of war, captured from a small trawler, were accommodated on mats on the deck of the after torpedo room. One prisoner had a small shrapnel wound in the right knee, wound was cleansed and debrided, sulfathiazole powder, one suture, and dressing applied, no complication or infection have appeared. All the prisoners appeared to be comfortable and extremely pleased with their surroundings. They have been helping with routine cleaning of the ship, and doing their own mess cooking. About half of the crew have been taking the multiple vitamin capsules provided.

(P) MILES STEAMED – FUEL USED

Pearl Harbor to Midway	1,219
Midway to Area	2,117
In Area Surface	808
Submerged	236
Area to Pearl	3,430

(Q) FUEL OIL EXPENDED

Midway to Area	23,890 gals.
In Area	5,062 gals.
From Area to Midway	31,781 gals.

---

* Pediculi Pubis – Crab lice.

U.S.S. *Wahoo* (SS-238)

(R) FACTORS OF ENDURANCE REMAINING

Torpedoes	14
Fuel	31,267 gals.
Provisions	40 days.
Personnel Factors	40 days.

(S) REMARKS

It is recommended that all influence torpedoes be provided all submarines going on patrol, and permit the individual submarines to inactivate the influence feature as necessary and as desired.

This will give the various submarines more flexibility in torpedo firing.

(Endorsements to sixth war patrol report)

FB5-42/A16-3   SUBMARINE DIVISION FORTY TWO
Serial   025   In Care of Fleet Post Office,
                     San Francisco, California,
                     4 September 1943.

CONFIDENTIAL

From: The Commander Submarine Division Forty-Two.
To: The Commander Submarine Force, Pacific Fleet.
Via: The Commander Submarine Squadron Four.
Subject: U.S.S. *Wahoo* – Report of Sixth War Patrol - Comments on.

     1. The *Wahoo* spent a total of twenty-eight days on this patrol. Of this time only seven days were actually spent in the assigned area. The patrol was terminated early after the expenditure of ten torpedoes on six targets without having inflicted any apparent damage on the enemy.
     2. This is the fourth patrol of the present Commanding Officer and the first patrol subsequent to a Navy Yard overhaul. The first three patrols of this Commanding Officer were outstanding and highly successful and resulted in the sinking and damaging of considerable tonnage.
     3. Nine separate torpedo attacks were made on six different targets. With the exception of the third attack in which a spread of two torpedoes was used all attacks were delivered with single torpedoes. The impact of the torpedo on the target fired on the second attack was heard by both sound operators and at the same time a plume of spray at the target was also seen by the periscope officer. Other than this "dud" no hits were made on any of the targets.

4. All attacks were delivered under cover of darkness after the target had been tracked by radar for periods varying between forth-five minutes to an hour and a half. Precise data were obtained on all targets and the Commanding Officer in each case skillfully maneuvered his boat to an excellent firing position for the initial attack. The ranges in all cases except one were under eleven hundred and fifty yards and in general small gyro angles (about five degrees) were used on the torpedoes. In an endeavor to obtain improved torpedo performance low power was used on two attacks without success. On still another attack a track of one hundred fifty degrees was accepted in preference to one nearer ninety in the hope that a glancing impact was the answer.*

5. A skillful commander with precise instruments for obtaining target data has a natural reluctance to waste torpedoes on slow speed, small, or even medium sized merchant vessels. Why use two torpedoes when one may easily finish the job? Considering all factors including torpedo performance it is essential that spreads be used to insure destruction of the target.

6. The port propeller shaft squealed at high speeds when running deep. Other than this the *Wahoo* was in excellent material condition. This will be investigated and remedied during refit. In view of the poor torpedo performance the torpedo tubes will be thoroughly and carefully checked while in dock and a special report submitted.

FC5-4/A16-3 SUBMARINE SQUADRON FOUR
Serial 0226 In Care of Fleet Post Office,
              San Francisco, California,
CONFIDENTIAL 6 September 1943.

FIRST ENDORSEMENT to
CSD 42 Conf. Ltr. FB5-42/A16-3
Serial 025 of 4 September 1943.

From: The Commander Submarine Squadron Four.
To: The Commander-in-Chief, United States Fleet.
Via: (1) The Commander Submarine Force, Pacific Fleet.
     (2) The Commander-in-Chief, Pacific Fleet.
Subject: U.S.S. *Wahoo* - Report of Sixth War Patrol – Comments on.

---

* This was actually a good idea, since an angled hit was somewhat more likely to explode. This patrol was conducted at about the same time that extensive (and technically unauthorized) tests finally provided the solution to the multiple problems with the Mark-14 torpedo and Mark-6 exploder.

U.S.S. *Wahoo* (SS-238)

1. The sixth war patrol of the *Wahoo* was conducted with the same aggressiveness which has made her past performances outstanding. Nine attacks were made within seven days on station, indicating a productive area.

2. All attacks were made from ideal positions, average torpedo run 1,070 yards, gyro angles small, all tracks close to 90°. Only one hit was observed and this was a dud. Accurate determination of the cause of the misses is of course impossible. One possible cause is the fact that all torpedoes were set for shallow depths, average depth setting 5.5 feet. The unreliable torpedo performance with shallow depth settings has been noted in the past. The decision of the commanding officer to fire single torpedoes, while understandable is not concurred in. A minimum of two, preferably three torpedoes, using a spread, should be fired at any target worthy of torpedo expenditure, taking into consideration the poor performance of the Mark XIV torpedo, the many unknowns in torpedo firing and the fact that medium size vessels can withstand one torpedo when it isn't in a vital spot.

3. The destruction of the sampans by gun fire was conducted with the usual efficiency of the *Wahoo*. It is recommended that the *Wahoo* be credited with the following damage inflicted on the enemy.

                              Sunk   3 Sampans.

Copy to:
  CSD 42
  CO WAHOO

FF12-10/A16-3(5)/(16)   SUBMARINE FORCE, PACIFIC FLEET
Serial   01235   In Care of Fleet Post Office,
                            San Francisco, California,
                            8 September 1943.

CONFIDENTIAL

THIRD ENDORSEMENT to
WAHOO Report of Sixth
War Patrol dated 8-29-43.

From: The Commander Submarine Force, Pacific Fleet.
To: The Commander-in-Chief, U. S. Fleet.
Via: The Commander-in-Chief, U. S. Pacific Fleet.
Subject: U.S.S. *Wahoo* (SS-238) – Report of Sixth War Patrol,
       (2 August to 29 August 1943).

Enclosure: (A) Copy of subject war patrol report.
          (B) Copy of Comsubdiv 42 Conf. 1st. end. FB5-42/A16-3

U.S.S. *Wahoo* (SS-238)

Serial 025 dated 4 September 1943.
(C) Copy of Comsubron 4 Conf. 2nd. end. FC5-4/A16-3,
Serial 0226 dated 6 September 1943.

1. The *Wahoo*'s sixth war patrol was the first after a navy yard overhaul. It was carried out in the Japan Sea.
2. Many contacts were made and nine aggressive torpedo attacks carried out. The lack of success of these attacks is being investigated. Failure to use torpedo spreads during most of the attacks undoubtedly contributed materially to the lack of success. Torpedo spreads must be used to cover possible errors in data or possibly of duds.
3. Three sampans were sunk by gunfire and six prisoners were brought back from one of them.
4. This patrol is not considered successful for Combat Insignia award.
5. The *Wahoo* is credited with inflicting the following damage to the enemy:

SUNK
3-Sampans - 96 tons

C. A. LOCKWOOD, Jr.

Patrol Seven, 9 Sep 1943 – 11 Oct 1943

Editor's Note:

Because *Wahoo* was lost during her seventh war patrol no official patrol report could be filed. The following is taken from *United States Submarine Losses, World War II*, issued by the Navy History Division, Office of the Chief of Naval Operations, 1963 edition.

WAHOO (SS-238)

Wahoo returned to Pearl Harbor from her sixth war patrol on 29 August 1943 with the dejected air peculiar to a highly successful submarine who suddenly could not make her torpedoes run true. In twenty-eight days away from port, seven of them spent in her assigned area in the Sea of Japan, *Wahoo* had expended ten torpedoes in nine attacks without inflicting any damage on the enemy. Her skipper, Cdr. D.W. Morton, returned to port to have the torpedoes changed or checked, and requested that *Wahoo* be sent back to the Sea of Japan for her seventh patrol.

On 9 September, *Wahoo* again departed Pearl. She topped off with fuel at Midway and left there on 13 September heading for the dangerous but important Japan Sea. Shortly afterwards, *Sawfish* left Midway and also headed for this area. *Wahoo* was to pass through Etorofu Strait, in the Kurile Islands, and La Pérouse Strait, between Hokkaido and Karafuto, and enter the Japan Sea about 20 September. She was to head south and remain below 43 degrees north after 23 September, and below 40 degrees north after 26 September. *Sawfish* was to follow *Wahoo*, entering the Japan Sea about 23 September and patrolling the area north of *Wahoo*.

No transmission was received from *Wahoo*, either by any shore station or by *Sawfish*, nor was she sighted by *Sawfish* after she left Midway. She had

U.S.S. *Wahoo* (SS-238)

orders to clear her area not later than sunset 21 October 1943, and to report by radio after passing through the Kurile Island chain en route to Midway. This report was expected about 23 October, but Midway waited in vain. By 30 October, apprehension was felt for *Wahoo*'s safety and an aircraft search along her expected course was arranged. When this revealed nothing, *Wahoo* was reported missing on 9 November 1943.

Although no transmission was received from *Wahoo* after her departure on patrol, the results of one of her attacks became known to the world via a Tokyo broadcast. Domei was quoted as reporting that on 5 October, a "steamer" was sunk by an American submarine off the west coast of Honshu near the Straits of Tsushima. It was said that the ship sank "after several seconds" with 544 people losing their lives. The submarine could have been none other than *Wahoo*; none other was operating in that area.

Information gleaned from Japanese sources since the cessation of hostilities indicates that an anti-submarine attack was made in La Pérouse Strait on 11 October 1943. This was two days after *Sawfish* went through the Straits. Supplementary data on the attack of 11 October state, "Our plane found a floating sub and attacked it, with 3 depth charges." *Sawfish* was attacked here while making her passage, and that attack is not mentioned in Japanese records; the primary attacking agency in that case was a patrol boat, and about five depth charges were dropped. Thus it is safe to assume that the attack cited here was made on *Wahoo*, and is not the attack on *Sawfish* with an incorrect date. Both Tsushima Straits, where the attack on the steamer was made, and La Pérouse Straits, through which *Wahoo* was to make good her exit from the Japan Sea, are known to have been mined. This despite the fact that *Sawfish* transited La Pérouse on 9 October and reported no indications of mining. It is felt, however, that *Wahoo* succumbed to the attack referred to above, and not to a mine.

Japanese records now reveal that the following ships were sunk in the Sea of Japan shortly before *Wahoo*'s loss: *Taiko Maru*, 2,958 T., 25 Sept.; *Konron Maru*, 7,903 T., 1 Oct.; *Kanko Maru*, 1,288 gt., 6 Oct.; and *Kanko Maru*, 2,995 gt., 9 Oct. *Wahoo* was the only submarine who could have sunk these ships.

Lost Aboard U.S.S. *Wahoo* (SS-238)

Name	Rate	Name	Rate
Anders, Floyd	MM3	Lape, Arthur D.	F1
Andrews, Joseph S.	EM1	Lindemann, Clarence A	S1
Bailey, Robert E.	SC3	Logue, Robert B	FC1
Bair, Arthur I.	TM3	Lynch, Walter L.	F1
Berg, Jimmie C.	MM3	MacAlman, Stuart E	PhM1
Brown, D.R.	ENS	MacGowen, Thomas J.K.	MoMM1
Browning, Chester R.	MoMM2	Magyar, Albert J.	MM3
Bruce, Clifford L.	MoMM1	Manalisay, Jesus C.	St3
Buckley, James P.	RM1	Mandjiak, Paul A.	MM3
Burgan, William W.	LT	Massa, Edward E.	S1
Campbell, J.S.	ENS	Maulding, Ernest C.	SM3

U.S.S. *Wahoo* (SS-238)

Carr, William J.	CGM	Maulding, George E.	TM3
Carter, James E.	RM2	McGill, Thomas J.	CMoMM
Davison, William E.	MoMM1	McGilton, Howard E.	TM3
Deaton, Lynwood N.	TM1	McSpadden, Donald J.	TM1
Erdey, Joseph S.	EM3	Mills, Max L.	RT1
Fiedler, E.F.	LTJG	Misch, G.A.	LTJG
Finkelstein, Oscar	TM3	Morton, Dudley W.	CDR – CO
Galli, Walter O.	TM3	Neel, Percy	TM2
Garmon, Cecil E.	MoMM2	O'Brien, Forest L.	EM1
Garrett, George C., Jr.	MoMM2	O'Neal, Roy L.	EM3
Gerlacher, Wesley L.	S2	Ostrander, Edwin E.	MM3
Goss, Richard P.	MoMM1	Phillips, Paul F.	SC1
Greene, Hiram M.	LT	Rennels, Juano L.	SC2
Hand, William R.	EM2	Renno, Henry	S1
Hartman, Leon M.	MM3	Seal, Enoch H., Jr.	TM2
Hayes, Dean M.	EM2	Simonetti, Alfred R.	SM2
Henderson, R.N.	LT	Skjonsby, Verne L.	LCDR – XO
Holmes, William H.	EM1	Smith, Donald O.	EM1
House, Van A.	S1	Stevens, George V.	MoMM2
Howe, Howard J.	EM3	Terrell, William C.	QM3
Jacobs, Olin	MoMM1	Thomas, William	S1
Jasa, Robert L.	MM3	Tyler, Ralph O.	TM3
Jayson, Juan O.	Ck3	Vidick, Joe	EM2
Johnson, Kindred B.	TM1	Wach, Ludwig J.	Cox
Keeter, Dalton C.	CMoMM	Waldron, Wilbur E.	RM3
Kemp, Wendell W.	GM1	Ware, Norman C.	CEM
Kessock, Paul	F1	Whipp, Kenneth L.	MM2
Kirk, Eugene T.	S1	White, William T.	Y2
Krebs, Paul H.	SM3	Whitting, Roy L.	MM3

Appendix I

Time Zone Indicators:

The wartime United States Navy designated each of the world's time zones using a letter. This table uses the wartime phonetic alphabet, which differs from the one used by the military today.

GMT   Zebra (Greenwich, England)
GMT +1 Able
GMT +2 Baker
GMT +3 Charlie
GMT +4 Dog
GMT +5 Easy
GMT +6 Fox
GMT +7 George
GMT +8 How
GMT +9 Jig
GMT +10 King
GMT +11 Love
GMT +12 Mike (West of dateline)
GMT −12 Yoke (East of dateline)
GMT −11 X-ray (Hawaiian time)
GMT −10 William
GMT −9 Victor
GMT −8 Uncle (U.S. Pacific time)
GMT −7 Tare (U.S. Mountain time)
GMT −6 Sierra (U.S. Central time)
GMT −5 Roger (U.S. Eastern time)
GMT −4 Queen (U.S. Atlantic time)

U.S.S. *Wahoo* (SS-238)

GMT −3 Peter
GMT −2 Oboe
GMT −1 Nan

Appendix II

Vessel Type Designations:

The United States Navy used letter codes to designate different classes of vessel. On American naval vessels, these codes were prefixed to the hull number and served to indicate both the class and actual designation of the ship. Submarines used the same codes to indicate target types, also applying the Navy designator to commercial vessels of the same type.

The designators were:

CV	Aircraft Carrier
CVL	Aircraft Carrier, Light[1]
CVE	Escort Aircraft Carrier[2]
AV	Seaplane Tender
BB	Battleship
CA	Heavy Cruiser[3]
CL	Light Cruiser
AO	Oiler[4]
AP	Transport[5]

---

[1] Built on converted cruiser hulls.

[2] Originally, these were converted from merchant hulls. Later CVEs were purpose built. In the Pacific CVEs were mainly used as aircraft ferries. In the Atlantic they were mainly used as anti-submarine units.

[3] Heavy Cruisers had a main armament larger than 6-inches (usually 8-inch). Light cruisers mounted 6-inch or smaller main guns. Both cruiser classes were armored, and designed for high-speed operations.

[4] Submariners applied the AO designation to commercial tankers, as well as Navy fleet oilers.

U.S.S. *Wahoo* (SS-238)

AK	Cargo Ship[6]
DD	Destroyer
DE	Destroyer Escort[7]
SS	Submarine
AM	Minesweeper
PC	Patrol Craft
PCE	Patrol Craft, Escort
SC	Sub Chaser
PT	PT Boat

---

[5] In military parlance, a transport is a troop carrier. Like the U.S. Navy (and Army), the Japanese used both purpose-built troop transports and converted liners in the role.

[6] General cargo ships, including civilian freighters.

[7] Destroyer Escorts (Frigates, in British parlance) were smaller than destroyers, and specialized for anti-submarine and escort duty. Most American DEs were powered by electric motors, driven by either diesel or steam-turbine generators, as this design was found to be both cheaper and quicker to build than conventional steam turbine propulsion.

Appendix III

Sinkings Credited, Wartime vs. JANAC

During the war, submarines were credited with the sinking of enemy ships based on patrol reports and intelligence sources. Following the war, when primary Japanese sources became available, the Joint Army Navy Assessment Committee (JANAC) was organized to compare claims with enemy documents. The results have been a source of controversy ever since.

American skippers have cursed the Japanese as poor record keepers, insisting that they *did* sink a ship on a particular day, despite lack of enemy documentation. In some cases, critical records were destroyed in bombing raids, or lost when the ships carrying them were also sunk. In other cases, of course, a ship claimed as sunk may have managed to limp away, or not have been damaged at all. (The latter was particularly problematic early in the war, when torpedoes were all too likely to explode prematurely, or not at all.)

Wahoo's wartime and JANAC credits are listed here.

Patrol	Captain	Wartime		JANAC	
		Number	Tonnage	Number	Tonnage
1	Kennedy	1	6,400	0	0
2	Kennedy	2	7,600	1	5,400
3	Morton	5	31,900	5	11,300
4	Morton	8	36,700	9	20,000
5	Morton	3	24,700	3	10,500
6	Morton	0	0	0	0
7	Morton	1	7,100	4	13,000

Printed in the United States
26947LVS00002B/82-99